Identification of Textile Materials

Seventh Edition

The Textile Institute
Manchester
1975

THE TEXTILE INSTITUTE
10 Blackfriars Street, Manchester, M3 5DR

© The Textile Institute, 1975

Third Edition, Revised and Enlarged, 1951
Third Edition, Second Impression, 1953
Third Edition, Third Impression, 1954
Third Edition, Fourth Impression, 1955
Third Edition, Fifth Impression, 1955
Fourth Edition, Revised and Enlarged, 1958
Fourth Edition, Second Impression, 1961
Fifth Edition, Revised and Enlarged, 1965
Fifth Edition, Reprinted with Corrections, 1967
Fifth Edition, Third Impression, 1968
Sixth Edition, Revised and Enlarged, 1970
Seventh Edition, Revised and Enlarged, 1975

ISBN 0 900739 18 5

PRINTED BY EYRE & SPOTTISWOODE LIMITED AT GROSVENOR PRESS, PORTSMOUTH

Identification of Textile Materials

The following Sub-committee of the Textbook Committee of the Textile Institute has contributed to the preparation of this seventh edition.

D. R. PERRY, J.P., MPHIL, FTI, MInstP, AMBIM, CGIA (Convener) (Blackburn College of Technology and Design)

H. M. APPLEYARD, FTI, DipRMS (International Wool Secretariat)

G. CARTRIDGE, ATI (ICI (Organics Division))

P. G. W. COBB, FRIC (Home Office Forensic Science Laboratory, Aldermaston)

G. E. COOP (ICI Fibres)

Miss B. LOMAS, MSc, ATI (University of Manchester Institute of Science and Technology)

G. G. RITCHIE, BSc, PhD (Formerly ICI Fibres)

C. TAYLOR, ATI (Tootal Ltd)

M. J. WELCH, BSc, ARIC (Courtaulds Ltd)

Staff

CAROLYN A. FARNFIELD, BSc, MSc, AIInfSc

Comments, suggestions for improvement, and communications for future inclusion should be sent to the General Secretary, The Textile Institute, 10 Blackfriars Street, Manchester, M3 5DR.

Contents

1 Introduction

Although no fundamentally new polymers have emerged for application in the textile field, some modified fibres produced by familiar polymer systems have been introduced. There is also a limited number of new polymer systems from which fibres are being produced for limited applications, such as high-temperature resistance. It will be appreciated, therefore, that it is necessary to keep the whole subject of fibre identification under continued revision and to make reference to any improved or alternative methods of examination, such as those developed to deal with bicomponent fibres. On the basis of these various aspects, the Sub-committee decided to adhere to the earlier preliminary method of classification of thermoplastic man-made fibres, using the simple tests for the presence or absence of the elements nitrogen and chlorine; these tests, suggested by W. J. Roff, permit the isolation of four main groups of fibres. Other groups of man-made fibres emerge from simple sorting tests involving the observation of non-thermoplastic behaviour (e.g., viscose) or high-temperature resistance (e.g., Nomex or Kevlar 49). This well tested approach has not been invalidated by the introduction of any major new fibre, although the detailed isolation of individual fibres within any one group has been the subject of considerable revision. The policy adopted for the Seventh Edition, therefore, has been to adhere to the main schematic approach used in earlier editions, but to include any new developments in identification techniques and also to place much more emphasis on the use of microscopy in the field of fibre identification.

The Sub-committee has considered carefully the terms of reference of this publication. While it is essentially a working document for the purpose of laboratory tests leading to the identification of an unknown fibre, other functions are evident, such as a document describing commercially available fibres and providing information on properties, and as an album of photomicrographs. In this latter respect we hope that the book will prove to be even more useful than the earlier editions. Thus although a certain duality of purpose appears to have evolved, the main term of reference of fibre identification has been retained. It should be stated, however, that no attempt has been made to deal comprehensively in this edition with methods for the quantitative analysis of fibre blends. The Sub-committee agreed that this topic was outside its terms of reference and recommends reference to the well-documented methods of analysis.

When defining terms of reference, it is necessary to state that the results of the tests concerned with man-made fibres relate to fibres that have not been modified by means of durable finishes, heat treatment, or dyestuffs that may significantly alter fibre behaviour. A serious identification problem exists with materials that have been modified, and the present day complexity of fibre-modifying finishes and dyestuffs introduces a formidable extension to the problem of identification. This topic is outside the scope of the present publication.

It should be noted that, in certain tables, the omission of results is intended to imply that a test is not diagnostic for that particular fibre.

New Fibres

The policy of the major fibre producers to extend the types of fibres within existing classes has been reflected in the present edition. Thus the new physical forms of polyolefins produced by the formation of fibres and yarns from sheet and embossed films are included, as are the more extensive ranges of polyamide and polyester fibres. The increase in the production of bicomponent fibre systems, that is systems that form hybrid fibres from two chemically different polymers, has also been covered in the appropriate sections. In this edition, as in the previous one, the acrylic fibres are dealt with at some length, and reference is made to the small differences in behaviour owing to the use of varying copolymer constituents and differing physical conditions of fibre formation. The elastomeric fibres have increased in number since the sixth edition and there have been some changes. The sections on viscose, acetate, and modacrylic fibres have been revised, and the material on metal-foil yarns has been updated to cover the more varied forms of construction now employed. High-performance fibres, such as carbon fibres, Kevlar 49, metallic, and inorganic fibres, are also dealt with at greater length to give a proportional representation of the commercial changes in these areas.

The degree to which information on new fibres could be included was to a large extent dependent on the co-operation of the fibre producers concerned. The great majority of firms co-operated by sending information supported by photomicrographs, for which the Sub-committee is extremely grateful.

Methods of Identification

Instrumental methods of analysis have gained considerable acceptance since the sixth edition was published and are, therefore, described in the book. All the techniques involve the use of expensive equipment of the type that cannot be readily available in all laboratories, but where it is proved that use of these techniques is beneficial, people can generally purchase time on instruments operated by other laboratories. The usefulness of these methods depends on the existence of a library of reference graphs or spectrograms prepared under standard conditions and on the ability to match the unknown sample rapidly and accurately against these known samples.

Instrumental methods of analysis provide reliable means of distinguishing not only between the main chemical types of fibre, but also between closely related fibres with only minor differences in chemical and physical properties, such as the now very numerous group of acrylic fibres. However, most laboratories will still rely on the simpler methods described in the main scheme of analysis, with initial microscopical examination being of paramount importance owing to the increase in multifibre samples.

The methods used in the scheme of analysis are not sufficient for chemical identification in the strict sense, though they serve to identify the main chemical types very reliably. The chemical differences between closely related, individually named members of a single type are generally undisclosed, or may be non-existent. In these circumstances the only available course is to measure a range of properties on each of the named individuals, in the hope that each individual will be unique in respect of one or a group of these properties.

Most new fibres when in commercial production are given proprietary names but not all names appearing in the literature represent distinct individuals; the same name, generally with a suffix or prefix, may represent different individuals, or a name may be transferred from one fibre to another without alteration at all. The fibres named and dealt with in this edition are believed to be in current production (1975) under the names given.

Fibre Properties

The section dealing with the properties of textile fibres has been extended and updated. The classification of man-made fibres used in this section is based on the internationally agreed standard nomenclature[1].

The section does not purport to give a complete account of the properties of fibres, the descriptions being mainly confined to those properties that are useful for fibre identification. Numerical values of densities, birefringence, and melting points, where appropriate, are given in Appendix B, although it must be emphasised that the latter figures can depend on the history of the fibres concerned. Dimensions of natural fibres are also given.

Fibre Blends

The sixth edition included a section on blend analysis from the point of view of solubility effects that lead to the identification of individual component fibres. In the present edition this feature is again included, but with the addition of a quantitative microscopical method, devised by a member of the Sub-committee (see Appendix A).

Photomicrographs

Many new photomicrographs are presented in this edition, illustrating the new fibres and including some replacements of photographs in the earlier edition. The microscopical techniques used in the preparation of the photomicrographs include normal light microscopy, polarization microscopy, and electron microscopy; the last two techniques being more extensively used now than they were at the time of the sixth edition. No attempt has been made to make the album complete, or to achieve a completely uniform presentation; the photomicrographs are intended to illustrate features that are significant for the identification of the fibre, and where such features are lacking, then it may well be that no photograph is included.

Acknowledgments

The Sub-committee very gratefully acknowledges the specialist help in the

---

[1]International Organization for Standardization. ISO/R 2076.

form of text and photomicrographs supplied by the following:

Dr. E. J. Smith, Pilkington Brothers Ltd, Ormskirk;

Dr. J. O. Warwicker, Shirley Institute, Manchester;

Mr. C. G. Jarman, Tropical Products Institute, London;

Mr. D. V. Badami and his colleagues, TBA Industrial Products Ltd, Rochdale;

Mr. J. W. C. Miller, Shell Chemicals UK Ltd, Manchester;

Mr. R. W. Foster, Monsanto Textiles Ltd, Leicester;

Mr. C. Kefford, Du Pont (UK) Ltd, Leicester;

British Nonwovens Manufacturers' Association, London;

Dr. A. Newton, University of Manchester Institute of Science and Technology;

Paper Industries Research Association, Leatherhead;

Lambeg Industrial Research Association, Lisburn;

Mr. B. Albinson, Lurex Co. Ltd, King's Lynn;

Dr. R. Jeffries, Shirley Institute, Manchester;

Mr. W. R. Beath and Mr. J. R. Collins, Courtaulds Ltd, London and Coventry;

British Standards Institution, London, for permission to reproduce part of BS Handbook 11, and PD 6469: 1973;

Elsevier Scientific Publishing Co., Amsterdam, and Dr. K. L. Loewenstein for permission to reproduce extracts from 'The Manufacturing Technology of Continuous Glass Fibres'.

Sub-committee members have been helped considerably by contributions given to them by their co-workers within the organizations they represent. It is these organizations that, on behalf of the Textile Institute, the Sub-committee wishes to thank, not only for their willingness to allow time to be spent on attendance at committee meetings, but for the use of laboratory facilities, access to information, and clerical assistance, without which the work of revision could not have been undertaken.

Finally, all the members of the Sub-committee wish to put on record their personal thanks to Ms. C. A. Farnfield of the Textile Institute for the effort that she has put into this book, for editorial work, for the way in which she has kept us up-to-date with all the essential documents, and above all for the patience that a secretary needs when dealing with a committee.

2 The Properties of Textile Fibres

2.1 Natural Fibres

2.1.1 Animal Fibres

2.1.1.1 Wool and Hair

Wool and hair are complex in structure and consist of a cortex of approximately spindle-shaped cells, which are normally not resolved in the optical microscope, surrounded by a cuticle of overlapping scales. The lower margin of each scale, (i.e., the part nearest to the root end of the fibre) is closely attached to the fibre, whereas the upper edge may project outwards to a degree that varies with the type of fibre. The degree of overlap affects the total thickness of the cuticle and this varies in fibres of different origin. This scale arrangement is responsible for the lower coefficient of friction when the fibre is rubbed from root to tip compared with that when rubbed in the reverse direction. This differential friction effect is a factor contributing to the well-known felting properties possessed in some degree by most wool and hair fibres. The pattern formed by the scales is a useful characteristic for identifying certain fibres, e.g., cashmere, but it is not a feature that can always be used for this purpose: a particular scale pattern is not necessarily confined to fibres from one type of animal. The scale pattern in some fibres may change from root to tip; it is better, therefore, to examine as long a length of fibre as possible.

In some fibres, the cortex encloses an inner hollow core, termed a medulla, which may be either unbroken or interrupted along the length of the fibre. This portion of the fibre consists of a series of cavities that are 'air-filled'. When a medullated fibre is examined in whole mount, the medulla appears black until the mountant penetrates the cavities when it becomes translucent and the structure of the medulla is then visible. The form of the medulla may vary among fibres from different species of animals and sometimes between fibres from the same species of animal and is often a useful characteristic in fibre identification.

The identification of animal fibres can be made only by careful microscopical observation of such features as cross-sectional outline, the pattern formed by the scale margins, the regularity of fibre diameter along the length of the individual fibres, the type of medulla, if present, the thickness of the cuticle, and the distribution of pigment granules. To observe these features, mounts of whole fibres in a medium of suitable refractive index are essential. Liquid paraffin is a good general-purpose mounting medium, its refractive index being such that the scale margins and other structural features may all be visible in the same preparation. If, however, the fibres have a very wide medulla, or are densely pigmented or dyed to a dark shade, much of the detail will be obscured and it will then be necessary to make a scale cast (see Section 4.1.4).

Some important features used for identification purposes are referred to in the following notes and illustrated in Figs. 1–36. For details of techniques and descriptions of animal fibres reference should be made to 'The Microscopy of Animal Textile Fibres' by A. B. Wildman and 'Guide to the

Identification of Animal Fibres' by H. M. Appleyard (both published by Wira, Leeds).

All animal fibres, other than silk, are composed of the same chemical substance, the sulphur-containing protein, keratin. They are thus unique amongst fibrous materials in having a sulphur content of the order of 2·5–5%. The presence of sulphur may be readily demonstrated by dissolving in boiling 5% caustic soda and adding lead acetate solution, when a brown coloration, owing to the formation of lead sulphide, develops.

Wool and hair fibres burn, forming a black mass and giving a characteristic smell. They are insoluble in acids in the cold, but readily dissolve when heated in solutions of caustic alkalis (5% caustic soda). With nitric acid they are stained to a yellow colour which turns to an orange shade when the fibres are subsequently treated with caustic alkali (Xanthoproteic reaction).

The structural characteristics of wool and hair fibres are liable to modification during textile finishing processes, e.g., by alkaline treatment, bleaching, or chlorination, especially when the treatment is severe. They may also be modified by heat, by exposure to light as during growth of the wool ('tippy' wool) or in garments during normal wear, by mechanical action or by attack by mildew and bacteria.

Chlorination causes a modification of the scale structure, the scale margins gradually becoming less prominent and, if the treatment is sufficiently severe, complete removal of scales may result. This change is accompanied by an increase in the ability of the fibres to take up dyes in the cold, which is the basis of the well-known Kiton Red G test[1].

Wool (Figs 1–13)

There are very marked differences in the characteristics of different breeds of sheep and these give rise to very large inter-sample variations in wool, for example, between a long wool of the Lincoln type and wool from a Merino. This variation has resulted in the grading of wool for trade purposes by means of the well-known quality numbers, which correlate in some degree with dimensional characteristics. When wool fibres have been processed it is not possible to say with confidence from what breed of sheep or country they originated because wool fibres from several breeds have the same microscopical appearance. There are four main types of wool fibre: fine, coarse, outercoat, and kemps. Each type has a characteristic scale pattern so that it is possible to distinguish between types, although it is not possible to assign a type to a breed of sheep; for example, fibres from Merinos and undercoat fibres from double-coated British mountain breeds have the same characteristic scale pattern, and kemps from any breed have the same scale pattern near the root and another pattern on the remainder of the fibre, both patterns being characteristic of wool kemps. Thus wool fibres vary according to the type of fibre rather than to the breed of sheep.

Wool fibres are generally white, but naturally pigmented fibres are present in certain wools. The pigment occurs in the form of granules, sometimes finely and evenly dispersed, and sometimes aggregated into streaky masses. The cross-sectional shape of fine wool fibres and those of medium thickness varies from oval to circular; in general, the finer the fibre the

---

[1]E. G. H. Carter and R. Consden. *J. Text. Inst.*, 1946, **37**, T227.

nearer the contour approaches to circular. Coarse fibres are more irregular in outline, particularly the outercoat fibres from mountain breeds (Fig. 9), and scale margins are prominent.

Merino wools are free from medullated fibres, but other types often contain a proportion of these fibres. In some types kemp fibres are also present. These are medullated fibres of a special kind in which the diameter of the medulla is so great in relation to the diameter of the fibre that the cortex exists only as an extremely narrow zone surrounding the medulla. Kemp fibres are usually chalky white, very coarse and brittle, have a long tapering tip and taper over a short distance to the root-end which, when the kemp has been shed from its follicle, usually ends in what appears to the naked eye to be a small bulb. Kemps may also be pigmented, sometimes reddish, sometimes black in colour; such kemps are to be found in wools from some mountain breeds, e.g., in Radnor wools where they are frequently termed 'red kemps'.

Fibres from the face, legs, axillary, and inguinal regions of sheep are sometimes found as contaminants in processed material[2].These fibres differ from the normal wool fibres in that they are regular in diameter with slightly projecting scales; they have a wide lattice-type medulla and may or may not be pigmented. Their scale margins may be either smooth or crenate or may have a transitional stage between the two, forming a pattern that is similar to that observed on some fibres from the common goat.

The appearance of wool fibres may be modified by various agents. Thus bacterial attack results in the appearance of longitudinal striations followed by splitting of the fibres longitudinally with the subsequent separation of cortical cells. This change, occurring at the end of a fibre, results in the appearance of a 'brush-end'. Severe chemical treatment is generally indicated by such changes as lifting and curling of the free ends of the scales, removal of scales, and swelling of the fibres.

It should be noted that there is a general relationship between fibre length, diameter, and crimp according to the type of wool, for example, wool fibres from long-haired sheep are comparatively coarse and have low crimp, sometimes described as curl, whereas fine wools are short and highly crimped.

Skin Wool (Figs 14, 15)

The term 'skin wool' is applied to wool removed from the skins of sheep slaughtered for mutton; the term 'pulled wool' is also used. These wools may be removed by a number of methods including the sweating process and the methods of painting the skin with sodium sulphide or lime; wool obtained by the last-named method is termed 'slipe wool'. In the sweating process the pelts are first soaked in water and then hung in ovens when, due to bacterial action, the fibre follicles are loosened in the skin so that the wool can be pulled off by hand. In the alternative methods, pastes of sodium sulphide or lime are painted on the flesh side of the skins that are then allowed to stand for a period of time. During this period the follicles are loosened and the wool can afterwards be pulled from the skins by hand. The wools are roughly sorted for quality during the pulling stage.

---

[2]H. M. Appleyard and M. E. Adèle Perkin. *J. Text. Inst.*, 1965, **56**, T45.

The processes of de-woolling skins affect the handle and spinning properties of the wool, so that the use of the wool may be somewhat limited. The root-ends of skin wool fibres are often in a softer and more plastic state than the rest of the fibre because they consist of cells that are not fully keratinized. During the pulling process the root-end cells are torn from their connexions with the cells in the follicle bulb and, when viewed under the microscope, many of the root-ends are seen to be tapered, some showing dark patches (Fig. 14). Samples of fibres pulled during the sulphide and lime processes may contain fewer tapered fibres than those fibres pulled during the sweating process, but they frequently contain a fairly high proportion of ends that have been damaged by the chemical treatment. The damaged fibres appear to be 'shrivelled' and are similar to fibres that have been subject to an alkali treatment (Fig. 15). The root-ends of skin wool are distinguishable from fleece and wool fibres, which are clean-cut.

Re-processed Wool

Re-used wool forms an appreciable proportion of the raw material used in some sections of the woollen industry, from which the rag, shoddy, and mungo trade is inseparable. The raw material comes largely from three sources, (i) old garments, either knitted or woven, (ii) unworn clippings from the garment-making industry, (iii) thread waste from the spinning and weaving sections of the textile industry.

The function of the shoddy industry is to convert this material into a fibrous form so that it can be processed on woollen carding and spinning machinery. This is done by mechanical disintegration of the rags ('grinding' or 'pulling') and the severe treatment necessary results in extensive breakage of the fibres.

Much of the raw material processed by the shoddy industry has already suffered fibre damage to an appreciable extent as the result of the initial processes of manufacture, such as bleaching, carbonizing, dyeing, and anti-shrink treatment, and as the result of subsequent usage as clothing, and this is accentuated by the mechanical treatment it receives on pulling. Consequently, some short and partially shredded, split, and brush-ended fibres are often found in material made from a blend containing a proportion of shoddy. Other distinctive features are the less prominent scales in comparison with those of virgin wool and the presence of a wide variety of dyed fibres in the same sample.

Mohair (Figs 16, 17)

Mohair is obtained from the angora goat and is generally white. The coat consists essentially of non-medullated fibres, but some fibres having a narrow, interrupted medulla are usually present; chalky white kemps may also be present. Mohair fibres are uniform in diameter along their length and have a smooth outline, their scale margins often being difficult to detect in profile. In this last respect, the fibres contrast sharply with wool with which they are often blended. In whole mounts, short longitudinal streaks can often be seen in the cortex. In cross-section, single mohair fibres show no special features of value for identification purposes. The cross-sections of the higher grades tend to be circular, whereas the lower grades contain a considerable number of fibres that are elliptical in outline.

Cashmere (Figs 18, 19)

The coat of the cashmere goat consists of long outercoat fibres covering an undercoat of very fine and soft fibres. These two types are separated by a special carding and combing process, the fine fibres being separated as noils. In colour they are greyish-white, fawn, or tan, pure white being very rare.

The fine fibres are non-medullated and uniform in diameter with the scale margins relatively far apart and not very prominent. These characteristics are important for identification purposes. In whole mounts, pigment granules may be seen, but these are sparsely distributed and may be completely absent in the lighter shades. Except in the very fine fibres, pigmented fibres show a clear peripheral zone free from pigment.

The coarse fibres are also uniform in diameter and have a continuous medulla, the diameter of which rarely exceeds half that of the fibre and is frequently less.

Common Goat Hair (Figs 20, 21)

The fleece of the common goat consists of coarse outercoat fibres and fine undercoat fibres; the proportion of these fibres varies considerably between the breeds of goat. In profile, the fine fibres have the appearance of sheep's wool. The coarse fibres are fairly regular in thickness and fairly smooth, and they may be medullated, the medulla being fragmental, interrupted, or continuous. The cross-sectional shape of the fine fibres is almost circular, and that of the coarse fibres is either circular or flattened. Coarse fibres may be pigmented, sometimes very densely, with large aggregates of pigment arranged in a radial pattern. An important recognition feature of the scale pattern of some coarse goat hair is a transitional scale pattern between one that is an irregular mosaic and one that is waved with crenate margins; at this stage, the scale margins are partly crenate and partly smooth, as seen in Fig. 20.

Camel Hair (Figs 22, 23)

The coat of the camel consists chiefly of two kinds of fibre, (i) coarse outer fibres up to a length of 380 mm, and (ii) the down or fine fibres of length 25–125 mm. The latter are separated as noil by combing and are often blended with fine wool. The very strong coarse fibres are made into tops for yarns spun on the worsted system. The finest qualities are light fawn in colour and the lower qualities rather darker.

All camel hairs are very regular in outline and uniform in diameter along their length. The scale edges project so very slightly from the body of the fibre that they can usually be seen only when mounted in a medium of refractive index considerably lower than that of keratin.

Coarse camel fibres range from nearly circular to ovoid in cross-sectional outline. The medulla is continuous and narrow in relation to the diameter of the fibre, rarely exceeding in diameter half that of the major axis of the fibre section. The pigment granules are mostly in the form of large aggregates that tend to be more thickly distributed towards the centre of the fibre, the region underlying the cuticle being often fairly free from pigment. In the more lightly coloured fine fibres, the pigment granules are fewer in number and distributed more widely. In profile the pigment in both fine and coarse fibres may appear streaky.

It is almost impossible to see the scale pattern on medium and coarse camel hair because of the pigment and medulla. Casts must be made in order to see the scale pattern (see Section 4.1.4).

Llama Fibres (Figs 24, 25)

The alpaca is the most numerous of the species of llama, the other species being the llama, guanaco, and vicuña; the first two are domesticated, whereas the latter two are wild. The colour of the coats of the alpaca and llama ranges from white, through several shades of fawn and brown, to black. By blending and combing these natural sorts, a wide range of natural coloured shades may be obtained. Of this group, the llama produces the coarsest fibres, followed by the alpaca, the guanaco, and finally the vicuña. Fibres of the vicuña are mainly very fine with a very soft handle, and the colour of the fibres ranges from golden to a deep rich brown, with a proportion of white fibres from part of the fleece. The fleece contains a proportion of beard hairs that are easily removed during processing prior to spinning.

The fibres are generally uniform in diameter along their length and have smooth profiles. In cross-section, the fine- and medium-diameter fibres are oval to circular in shape and many of them have a relatively narrow medulla. In some coarse fibres the medulla may be bi-partite or multi-partite according to the contour of the cross-section; these features are characteristic of the coarse fibre from all the llama family. Some fine alpaca fibres are non-medullated but the majority have a medulla that may be either continuous or interrupted. The distribution of pigment in the fibres can be a useful diagnostic feature; in some fibres pigment is distributed sparsely and in others it is dense and often in large aggregates. Frequently it is most dense near the cuticle, accentuating its thickness; a thick cuticle may also be seen in the non-pigmented fibres.

Rabbit and Hare Fibres (Figs 26–28)

The coat consists of two main types of fibre, the coarse beard or guard fibres and the finer fur fibres. The guard fibres, sometimes referred to as shield fibres, have two portions, the shaft or proximal portion, which may be fine, and the distal portion, which is flattened and shield-like. These fibres are generally longer than the fur fibres and so project above the undercoat. The shield portions are often broken off when the fibres are being removed from the skins and are sold as 'rabbit kemps' to the textile industry. The fur is used in the fur-felt hat trade.

The medulla is like a ladder in appearance and in the shaft portion of the finer guard fibres it is uni-serial (a single series of cavities), but in coarser guard fibres the shaft may have a bi-serial medulla (double series of cavities). In both coarse and fine guard fibres, the medulla in the shield portion is multi-serial. Sections across the shields are frequently dumb-bell shaped, but may also be bean-shaped or elliptical, the medullary spaces being clearly seen. The finer fur fibres have no shield portion and the medulla consists of a single column of spaces. Their cross-sectional outline is markedly angular. In blends with wool, rabbit fibres are easily distinguished in cross-section.

The fibres of the angora rabbit are longer than those of the common rabbit and are often blended with wool in hosiery yarns. In general, their characteristics are similar to those of the common rabbit, the longer fibres having a shield-shaped portion that tapers into a thinner shaft portion below. The shield fibres vary in thickness and length and in the size of their shields. Shorter fur fibres, rather longer than those of the common rabbit, but having similar characteristics, are also found.

Horse Hair (Figs 29–31)

The two chief commercial types of horse hair, i.e., mane hair and tail hair, can be distinguished by microscopical examination. Apart from the evidence of 'flagging' (mechanical splitting of the tips when the hair is to be used for brushes), whole mounts give no details of diagnostic value, but the greater average thickness of tail hair can be noted. There are also no marked differences between the cuticular-scale patterns of mane and tail hairs.

From cross-sections, several distinct types of fibre differing in the size and the form of the medulla can be distinguished. Some types preponderate in mane hair, others in tail hair. The fibres are either circular or ovoid in cross-sectional outline, the latter predominating. Mane hair usually contains more circular fibres than tail hair. Most, but not all, horse-hair fibres are pigmented, a characteristic feature of all types being the tendency for the pigmentation to concentrate in the part of the cortex near to the medulla.

Cow Hair (Figs 32–34)

There are two commercial types of cow hair, i.e., body hair and tail hair. Body hair is relatively short whereas tail hair is long and often unpigmented. When pigment is present it is usually found in a band midway between the medulla and cuticle.

In body hair, the degree of pigmentation varies according to colour, but unpigmented fibres are usually present in coloured samples. Usually, a large proportion of fibres are unmedullated, even the coarse fibres having only a relatively narrow unbroken medulla.

White calf-body fibres are frequently found. These have long slender tips and almost all, both fine and coarse, are medullated. In the coarse fibres, the medulla, in cross-section, is usually concentric with the section outline and is unbroken, but it is often interrupted (and fragmental) in the finer fibres. In cross-section, the fibres are elliptical.

The tail fibres are very coarse but frequently unmedullated. The medulla, if present, is often very narrow and interrupted (and fragmental). In cross-section, the fibres are largely circular and the boundaries of the cortical cells can be clearly seen.

Yak (Figs 35, 36)

Yak hair is obtained from an animal of the cattle family which is native to parts of Asia. The fleece consists of long, coarse fibres with an undercoat of down fibres. Its colour varies from piebald to reddish-brown or black. According to Burns, Von Bergen, and Young[3], the fleece contains about

---

[3] R. H. Burns, W. Von Bergen, and S. S. Young. *J. Text. Inst.*, 1972, **53** (2), T45.

20% of down fibres. The average length of the coarse hair is about 160 mm and of the down fibres about 70 mm. The mean fibre diameter of the coarse fibres is about 63 μm and of the fine fibres about 23 μm.

In whole mount the coarse fibres are seen to be uniform in thickness and have smooth profiles with scale margins which protrude only slightly. The medulla, which is usually narrow or discontinuous, may be difficult to see because of the denseness of the pigment granules. The cross-sectional shape is oval to round with a fairly thick cuticle and, when present, the medulla is narrow. The pigment distribution varies from none to very dense and is usually evenly distributed. The scale pattern is crenate with near to close margins with no variation from root to tip.

The fine fibres are fairly regular in thickness with slightly prominent scale margins, they have no medulla and have very sparse to dense pigmentation. In cross-section they are oval to circular with a thin cuticle. The pigment distribution, which may be sparse to dense, may be arranged bi-laterally. The scale pattern near the root and along the length is waved mosaic with mostly smooth margins, a few margins are slightly crenate; near the tip of the fibres the pattern changes to waved crenate with near margins.

The hair is used locally for blankets. When the coarse and fine fibres are separated the former are used in carpet blends, the uses of the down fibres are similar to those for cashmere.

2.1.1.2   Silk

Silk from *Bombyx mori* (Figs 37–42)

Silk occurs in the cocoon as two filaments (brins) of the protein fibroin, coated and cemented together by a second protein, sericin, termed 'silk gum'. The raw silk yarns used in textile manufacture are made up of several twin continuous filaments, ranging in the more common yarns from five pairs up to eight or ten pairs of brins; these are processed into nett silk yarns by twisting and folding. Nett silk is usually woven or knitted before the gum or sericin, which acts as a natural size, is removed. The residue from the cocoons after reeling off the portion used for nett silk is collected, together with damaged and unreelable cocoons, and becomes waste silk. This waste silk, containing gum and other foreign matter, is made into thick, virtually twistless, strands, cut into lengths, boiled-off (degummed), and submitted to the 'dressing' process. The fibre lengths in the various drafts range from about 50 to 152 mm. These are spun into several grades of silk. The shortest fibres, the silk noils, are spun on the woollen system. Spun silk is always in the degummed state. Schappe silk, formerly spun from wastes that had been degummed by a natural fermentation process, contains up to 5% residual gum and has a rather wiry handle, whereas English spun silk, free from gum, is soft in handle.

Sericin differs slightly from fibroin in dyeing properties, and its presence can be shown by staining reactions[4]; under the microscope it can be seen adhering to the filaments and binding them together; it is removed by dilute alkali, and the normal industrial degumming process uses hot soap solution.

---

[4] e.g., using Colour Index Acid Blue 43.

Degummed silk filaments are fine, uniform, and without visible internal structure; in cross-section, they approximate to equilateral triangles with rounded apices. Under the action of severe rubbing or flexing, the filaments tend to show internal striations due to splitting into fine fibrils. Sometimes, in hanks of dyed silk, small light-coloured 'neps' can be seen on the yarn surface. Examination under the light microscope shows that these 'neps' are composed of tangles of fine fibrillar strands of silk of much finer linear density than the normal silk filaments. This defect is known as 'lousy silk' (see Figs 40, 41).

Silk swells and dissolves in cold concentrated hydrochloric acid, and in cold 80% sulphuric acid. It turns yellow, swells, and disintegrates in concentrated nitric acid. Its behaviour with hydrochloric acid distinguishes silk from wool, hair, and regenerated protein fibres. Silk gelatinizes quickly and dissolves in boiling 5% caustic soda, but not so readily as wool.

In weighted silk the weight lost in the degumming process is made up by other substances. The two main processes are (i) tin phosphate–silicate, in which the silk is soaked alternately in stannic chloride solution and sodium phosphate solution until the desired increase in weight owing to stannic phosphate has occurred, followed by a treatment in sodium silicate, and (ii) logwood, used only for blacks, in which the weight is added by treatment with logwood extract and metallic salts. Tin-weighted silk is distinguished from unweighted by the white ash skeleton, which becomes incandescent in the Bunsen flame. Logwood weighting is recognized by the red colour of both fibre and solution when the former is treated with concentrated hydrochloric acid.

Tussah Silk (Figs 43, 44)

This wild silk is of brownish colour, but it can be bleached to a fawn or cream shade. The individual filaments are coarser than those of *Bombyx* silk; in longitudinal view they are ribbon-like and show a striated and granular structure. The cross-sections have a markedly granular interior and in shape some approximate to elongated triangles.

Tussah silk resembles *Bombyx* silk in its properties, but is more resistant to acids and strong alkalis. It swells, but does not dissolve, in cold, concentrated hydrochloric acid; in boiling 5% caustic soda it disintegrates to a pulp without dissolving completely.

Tussah silk is not usually weighted.

Anaphe Silk (Figs 45, 46)

Anaphe silk is a wild silk and is never reeled as a continuous-filament or nett silk. It is used only as a waste silk for spinning. When examined under a microscope before degumming the two individual filaments, which are cemented together, appear as transparent rods with a line up the centre. The fibre is characterized by transverse lines across it at frequent intervals. The cross-section is crescent-shaped, each filament comprising half the crescent. Anaphe is difficult to degum, since it is necessary to boil it in a solution of soda ash, and most of the natural brown colour is removed with the gum. In resistance to chemical attack and general properties, Anaphe resembles Tussah rather than cultivated silk (*Bombyx mori*).

2.1.2 Vegetable Fibres

2.1.2.1 Seed Fibres

Cotton (Figs 47–50)

The cotton fibre is a seed hair formed by the elongation of a single epidermal cell of the cotton seed. This cell continues its growth in length with only a fine skin or cuticle for a cell wall until the maximum length of the hair is almost attained. The next stage in the development is the thickening of the wall; layers of cellulose are laid down on the inner side of the cuticle. When the boll ripens it bursts, the hairs dry, collapse, twist upon themselves, and form convoluted flattened tubes open at the base and tapering to a closed tip. This is the final form of the cotton fibre. The convolutions reverse their direction at intervals so that they are present in the same fibre in both clockwise and counter-clockwise directions in about equal proportions.

Cotton is distinguished from all other important textile fibres in that between crossed polars under the microscope it remains substantially bright in all orientations, except that in the orthogonal positions dark bands cross the fibre at frequent intervals indicating the reversal points of the underlying spiral structure.

The fibres in any one sample of cotton are not uniform in length or diameter, but average values for staple length and fineness can be defined by suitable procedures and serve to classify cottons for particular purposes. In general, long cottons are fine and short cottons are coarse. Fine cottons have more convolutions than coarse ones. The range of values of length and apparent diameter for the main classes of cotton are given in Table B1.2 (Appendix B).

Chemical and finishing treatments may have a marked effect on the appearance and properties of the fibre. Cotton is essentially cellulose, but in the raw state (grey cotton) it contains small quantities of fat and wax, pectin, proteins, and natural colouring matter. The removal of these impurities is effected by appropriate methods of scouring, kier boiling, and bleaching.

Lustre is improved by the mercerization process in which cotton materials are treated, usually under tension, with cold, concentrated solutions of caustic soda. This process causes the fibres to swell and lose most of their convolutions and leads to an increase in lustre. In any sample of mercerized cotton, considerable differences may be found in the degree of swelling and convolution of individual hairs.

Cotton fabrics may be treated with resins or other chemical finishes in order to improve dimensional stability and resistance to creasing, and to obtain glazed and embossed effects by mechanical after-treatment. Finishing processes involving chemical modification of the fibre include acetylation and cyanoethylation. Cotton is acetylated to improve its resistance to heat and microbiological attack, and to reduce its affinity for direct dyes.

Akund (Fig. 51)

Akund or 'Calotropis floss' is the seed hair obtained from *Calotropis procera* (Willd.) R.Br., and *C.giganta* (Willd.) R.Br., members of the botanical family Asclepiadaceae. The fibre is sometimes referred to as 'vegetable floss or silk'. It is fine, soft and lustrous, but very weak and hence, like kapok, it is rarely used by itself as a textile fibre.

Under the microscope the hairs appear similar to those of kapok but the base does not show the net-like thickenings seen in kapok (see Fig. 52). The average length ranges from 30 to 40 mm and the average diameter is 20 μm.

Kapok (Figs 52–54)

These fibres are unicellular hairs occurring in the pods that constitute the fruit of the plant; they grow on the inner wall of the pod and very occasionally on the seeds. There are two sources of kapok, both obtained from trees belonging to the botanical family Bombaceae; Java kapok is obtained from *Eriodendron anfractuosum* D.C., formerly known as *Ceiba pentandra* (L.) Gartn. Indian kapok comes from *Bombax malabaricum* D.C.

The fibre is smooth, cylindrical, hollow, thin-walled, and frequently bent over on itself. It tapers to a point at one end and the other forms a slightly bulbous base with annular or reticulate markings.

Owing to the smoothness and brittleness of the fibre walls, kapok is seldom used by itself as a textile fibre, but it may be mixed with cotton or wool and spun into a yarn. It is generally in the form of loose fibres that have been used for stuffing cushions, mattresses, and life jackets. The apparent density of the fibres is less than one-quarter that of water, but the apparent density of a loose mass may be between one-tenth and one-twentieth that of water because of the large amount of air entrapped between fibres. The walls of kapok fibres consist of highly lignified cellulose that may be detected by the usual reagents, e.g., zinc chlor-iodide (see Section 5.17.16). The average length of Java and Indian kapok ranges from 20 to 30 mm and the average diameter is 20 μm.

2.1.2.2 Bast Fibres[5,6]

Botanically, the term 'bast' is synonymous with phloem, the food-conducting tissue of vascular plants. In a broad commercial sense, the term 'bast fibre' is used to denote fibres obtained from the cortex and pericycle in addition to the phloem. These bast fibres that are obtained from the stems of various dicotyledonous plants are also often referred to in the trade as 'soft' fibres, to distinguish them from the leaf, or 'hard', fibres obtained from monocotyledons. The bast fibres of commerce are not discrete entities, but are composed of elongated thick-walled cells (ultimates) 'cemented' together both end to end and side by side, and forming bundles of filaments along the length of the stem.

The chief bast fibres of commerce are flax (*Linum usitatissimum*), jute (*Corchorus capsularis* and *Corchorus olitorius*), hemp (*Cannabis sativa*), and those fibres derived from *Hibiscus cannabinus* (kenaf, bimli, mesta, etc.) and *Hibiscus sabdariffa* var. *altissima* (roselle). Fibres from *Urena lobata* (urena fibre), *Crotalaria juncea* (sunn fibre), *Boehmeria nivea* (ramie), and other dicotyledons are also employed.

Care must be taken when the various trivial names of fibres are used[7]. The term 'hemp' is often loosely and erroneously appended to fibres

[5]D. Catling. 'Microscopy and Identification of Vegetable Fibres', H.M.S.O., *to be published 1976*.
[6]R. H. Kirby. 'Vegetable Fibres', Leonard Hill (Books) Ltd, 1963.
[7]H. R. Mauersberger. 'Matthews' Textile Fibers', Chapman & Hall, 6th edition, 1954.

obtained from plants other than *Cannabis sativa*, giving rise to such names as sunn 'hemp' (*Crotalaria juncea*), Deccan 'hemp' (*Hibiscus cannabinus*), Calcutta 'hemp' (*Corchorus capsularis*), bowstring 'hemp' (*Sansevieria* sp.), Manila 'hemp' (*Musa textilis*), and Mauritius 'hemp' (*Furcraea gigantea*). Similarly the term 'jute' is often misused (see under *Hibiscus* and *Urena* fibre).

The annual tonnage of jute fibre produced exceeds that of all the other bast fibres combined; *Hibiscus* fibres and flax come next, followed by hemp, sunn fibre, and *Urena lobata*. The production of other bast fibres is relatively very small, and their processing is in many cases carried out merely on an indigenous basis, solely for local uses.

The bast fibres are usually separated from the plant-stem tissues by retting, which consists in subjecting bundles of stems to controlled rotting in water for a period dependent on the particular conditions and fibre type concerned. The separation is effected by the action of bacteria that primarily attack the parenchymal tissues in which the fibre strands are embedded, and the retting may be carried out in ditches, dams, rivers, etc., or in specially constructed tanks, under carefully regulated conditions. Chemical separation of the fibres is also occasionally employed, and mechanical isolation and stripping, either alone or in combination with some form of retting, is sometimes used.

Flax (Figs 55–57)

Flax is obtained from the stem of the flax plant (*Linum usitatissimum* L.). The fibre is produced mainly in temperate climates; the Soviet Union being by far the largest producer. The number of fibre bundles in the stem ranges from 15 to 40, and each bundle contains from 12 to 40 ultimate fibres. After retting, the fibre bundles are separated from the cortex and woody tissue by scutching, which is a mechanical process. A certain amount of cortical tissue is left adhering to the fibre bundles and gives the characteristic colour to the raw flax strands.

The ultimate fibres consist of pointed cells with very thick walls and very small lumina. A peculiarity of the flax fibre, in common with the majority of bast fibres, is the presence of transverse dislocations, often in the form of an X, which show up very clearly when the fibres are mounted in liquid paraffin. The walls of the fibres have a spiral fibrillar structure, the external fibrillae running in a direction corresponding to an S twist, and this is the basis of the drying-twist test for the identification of flax (see Section 5.12). 'Cottonized flax' is the name given to a cotton-like material made by the separation of flax into its ultimate fibres by treatment with alkali. Descriptions of the microscopy of flax are given by Osborne[8] and by Slattery[9].

Jute (Figs 58–60, 79(a))

Jute fibre is obtained from the bast of two plants, *Corchorus capsularis* and *Corchorus olitorius*, of the botanical family Tiliaceae. The main source of commercial supply for the world market is Bangladesh, although India produces an equivalent quantity but for internal use only. Other countries,

---

[8]G. G. Osborne. *Text. Res.*, 1935, **5**, 356; 431.
[9]E. Slattery. *J. Text. Inst.*, 1936, **27**, T101.

such as China, Burma, Brazil, Nepal, and Thailand, also grow jute but in small quantities only.

The fibre lies along the length of the plant stem in the form of an annular meshwork composed of more than one fibre layer. The commercial fibres, as obtained from the plant, are 1·5–3 m long, and, when viewed in transverse section under the microscope, show from 6 to 20, or even as many as 50, single thick-walled polygonal cells (ultimates) each containing a central canal or lumen; in places, the lumen broadens considerably and the walls are correspondingly thin. In longitudinal view, the tips of the ultimates are seen to be pointed.

The relatively high lignin content of jute, as shown in the phloroglucinol test (see Section 5.11) distinguishes the fibre from flax and hemp, quite apart from the differences in the lengths of the fibre ultimates. A general review of the jute fibre structure has been given by Parsons[10], and a more detailed study of the microscopy of jute fibre has been presented by Osborne[11].

Hibiscus **and** *Urena* **Fibre (Fig. 79(b))**

Owing to the shortages of jute that occurred during World War II and on several occasions in the post-war period, great interest had arisen in jute-like fibres and their production, with a view to supplementing the supplies of jute available. As a result of this interest, some fibres are now recognized commercially in their own right, in particular, those fibres derived from plants of *Hibiscus* species.

The two most important plants for fibre production are *Hibiscus cannabinus* and *Hibiscus sabdariffa* var. *altissima*; fibre from both these plants is called 'kenaf'. Other common names for *H. cannabinus* are 'mesta' and 'bimli jute' in India and 'Java jute' in Indonesia. *H. sabdariffa* var. *altissima* is called 'Siam jute' in Thailand, 'mesta' in India and Pakistan, 'roselle' in Indonesia, and 'ke-nap' in Vietnam. Although much effort has been expended on trials with *H. cannabinus*, the world production is much less than that of the less publicized *H. sabdariffa*.

'Congo jute', as it is known, is obtained from the plant *Urena lobata* L. Commercially, this fibre is much less important, and output is comparatively small, coming mainly from the Congo region. It is also grown in Brazil, West Africa, and Madagascar. The dimensions and chemical and physical properties of *Urena* fibre are very similar to those of kenaf, and, like the *Hibiscus* species mentioned above, *Urena lobata* is a member of the Malvaceae.

The Malvaceae are closely related to the Tiliaceae, to which belongs the true jute plant, and the stem anatomy is similar in both families. This similarity extends to the microscopical details and chemical composition of the fibre bundle and the ultimates. Physical properties, too, are similar to those of jute, although it may be said that, on average, kenaf fibre is coarser and slightly more extensible than jute.

However, Soutar and Bryden[12] have reported a method of differentiating jute from *Hibiscus* fibres or *Urena lobata*, based on differences in the

[10]H. L. Parsons. 'Jute', *Textile Institute Handbook of Textile Technology,* No. 4, 1949.
[11]G. G. Osborne. *Text. Res.,* 1935, **5**, 362; 437.
[12]T. H. Soutar and M. Bryden. *J. Text. Inst.,* 1955, **46**, T521.

acetyl content of the fibres, but the method cannot be used if the fibres have been treated with alkali, since this readily releases the acetyl groups. Jarman and Kirby[13] have published details of a method of differentiating jute from *Hibiscus* fibres or *Urena lobata*, based on differences in the types of crystals found in the ash of the fibre. This method is most successful when applied to the raw fibre, since the crystals are associated with extraneous matter that is removed during processing.

Hemp (Figs 62, 65(a), 80(a))

Common or true hemp is derived from the stem of *Cannabis sativa* and is closely allied to flax as it, too, is a product of temperate climates, the main producing areas being the Soviet Union and Eastern Europe. The ultimate fibres are not unlike those of flax, both in their dimensions and general appearance. The external fibrillae of the cell wall run, however, in a direction corresponding to a Z twist, giving a counter-clockwise twist as opposed to a clockwise one as given by flax in the drying-twist test (see Section 5.12). The presence of crystals in the ash of hemp, except with boiled or bleached fibre, is another distinguishing feature. Kundu and Preston[14] describe the fine structure of the fibre.

Sunn Fibre (Figs 63, 64, 65(b))

This fibre, which is usually referred to as sunn hemp, is produced commercially almost entirely in India. It is obtained from the plant *Crotalaria juncea* L., a member of the Leguminoseae. The fibre resembles true hemp in appearance and has been used as an alternative to hemp in the making of twines. It is also used in the manufacture of cigarette papers. Under the microscope the ultimates appear somewhat similar to those of true hemp but tend to be shorter; however this difference cannot be relied upon with certainty. Differences, such as the presence of crystals associated with hemp (see Fig. 80(*a*)) but not with sunn fibre, and in the epidermis (see Fig. 65), are useful aids to identification. Klenk[15] gives a detailed description of the distinguishing features of hemp and sunn fibre.

Ramie (Figs 66, 67)

Ramie is obtained from the stems of *Boehmeria nivea* Gaud., especially the variety *tenacissima*, belonging to the Urticaceae or nettle family. It has been grown in China for centuries. Other producers include Brazil, Colombia, Japan, and India. The stems contain a high proportion of gums and pectins which necessitate special methods of preparation. The first stage consists in stripping the ribbons of fibre from the stem; this may be done either mechanically or by hand after immersing the stems in water. The fibre reaches European markets in the form of these ribbons, which are known as 'China grass'. Some form of degumming treatment, such as an alkali boil, is necessary to separate this 'China grass' into fibres for spinning.

[13]C. G. Jarman and R. H. Kirby. *Colon. Pl. Anim. Prod.*, 1955, **5**, 281.
[14]B. C. Kundu and R. D. Preston. *Proc. Roy. Soc. B*, 1940, **128**, 214.
[15]H. Klenk. *Textil-Praxis*, 1962, **17**, 432.

The fibre is characterized by the exceptionally long ultimates. These have a mean length of about 150 mm and vary in diameter from 25 to 75 μm. They are thick-walled and flattened in cross-section, and there is considerable variation in the cross-sectional area of the individual ultimate. Along the ultimates, cross-markings and longitudinal striations are of frequent occurrence. The microscopy of ramie has been described by Osborne[16].

2.1.2.3 Leaf Fibres

Leaf fibres are obtained from the leaves or leaf stalks of various monocotyledonous plants. Such fibres are usually associated with the vascular bundles, traces of which, in the form of spiral elements, are frequently visible on microscopical examination of the fibre. The leaf fibres are extracted by mechanical methods, and removal of the associated parenchyma cells is thus often less complete than is the case with bast fibres.

Sisal (Figs 68–70, 80(b))

Sisal[17] is, strictly, obtained only from the leaves of *Agave sisalana* Perrine, a plant native to Central America, but now cultivated as a fibre source in other parts of the American continent and in Africa and Asia. Its chief use is in the production of twines, cordage and ropes, sacking, carpeting, and other coarse fabrics. It is also used in papermaking. The fibres from other *Agave* plants, and particularly henequen (*Agave fourcroydes* Lemaire), resemble sisal very closely, and indeed are sometimes termed sisal. Fibre from the high yielding *Agave* hybrid No. 11648 now accounts for a substantial proportion of exports from Tanzania. The fibre is similar to that from *A. sisalana*, however, the fibre from the 'first cut' leaves is finer.

Sisal is distinguished from abaca by Billinghame's test (see Section 5.10), and the presence of rod-like crystals in the ash (Fig. 80(*b*)). It differs from *Phormium* fibre in that, when viewed in transverse sections, the ultimates are polygonal in outline with a rounded polygonal lumen. Nutman[18] describes the morphology and histology of the *Agave* fibres.

Abaca (Manila) (Figs 71, 72, 81(a))

This fibre is obtained from the leaf sheaths of the abaca plant (*Musa textilis* Nee.), which is cultivated principally in the Philippine Islands and also in Ecuador; its main uses are for cordage and papermaking. The fibre is pale cream and lustrous in appearance. Billinghame's test is commonly used for differentiating abaca (manila) from other cordage fibres. When the ash is examined under the microscope, the characteristic stegmata can be seen (see Fig. 81(*a*)). Stegmata are small silica cells that are present in the living plant in longitudinal files adjacent to the fibres. They vary in shape (as can be seen from Fig. 81) and are therefore a useful guide in identification. For a detailed description of the microscopy of the fibre, see Osborne [19].

---

[16]G. G. Osborne. *Text. Res.,* 1935, **5**, 75.
[17]G. W. Lock. 'Sisal', Longmans, 1962.
[18]F. J. Nutman. *Emp. J. Exp. Agric.,* 1937, **5**, 75.
[19]C. G. Osborne. *Text. Research,* 1935, **5**, 360; 435.

Phormium Fibre (Figs 73, 74)

This fibre is commonly called New Zealand flax or New Zealand hemp, although it appears that it is now produced only in South America and South Africa. It is not, however, related to either flax or true hemp, but is obtained from the leaves of *Phormium tenax* Forst. It is extracted mechanically, and hand-prepared fibre can be woven into a cloth resembling linen, although its main use is for the manufacture of twine and sacking. Microscopically, this fibre resembles sisal, but, in cross-section, the ultimates are mainly round or polygonal with a small lumen (compare Figs 70(*b*) and 74(*b*)).

2.1.2.4 Fruit Fibre

Coir (Figs 75–78, 81(b))

Coir is the reddish-brown fibrous mass contained between the outer husk of the coconut (the fruit of the palm *Cocos nucifera* L.) and the shell of the inner kernel. The fibre is obtained by steeping the husks in water for some time and then tearing off the fibre. The principal sources of supply are India and Sri Lanka. The best quality fibre is used for the manufacture of ropes and mats. The coarser and shorter fibres are employed as stuffings for mattresses. Coarse long fibres are used in brushmaking. Curled fibre is used in rubberized coir car seats and mattresses.

The ultimate fibres are short (less than 1 mm long) and the edges of some of the fibres have a distinct wavy outline. The round stegmata (see Figs 77, 81(b)) to be found in the ash are particularly characteristic. An account of the structure of the coir fibre, including the identification of chemical and biological damage using stains and swelling reagents, is given by Sen Gupta, Saxena, and Mukherjee[20].

2.1.3 Mineral Fibres

Asbestos[21] (Figs 82–88)

Asbestos is a generic term covering a number of naturally occurring fibrous, crystalline, inorganic silicates. These asbestiform minerals can be divided into two groups which are classified mineralogically as serpentine (sheet silicates) and amphibole (chain silicates).

The principal member of the serpentine family is chrysotile, a hydrated magnesium silicate whose composition approximates to the formula $Mg_3(Si_2O_5)(OH)_4$. Chrysotile accounts for approximately 90% of the world usage of asbestos. Electron microscopy has shown that the silicate sheets form scrolls and cylinders which are referred to as fibrils (Figs 84, 85) and very large numbers of these together form the fibre bundles visible to the naked eye. Individual fibrils have diameters in the range 300–350 Å and may be several millimetres in length resulting in aspect ratios as high as 10^5 which, coupled with their strength and flexibility, lead to their

---

[20]S. R. Sen Gupta, B. B. L. Saxena, and A. N. Mukherjee. *J. Sci. & Ind. Res. (India)*, 1949, **8B**, 61.

[21]C. Z. Carroll-Porczynski. 'Asbestos', Textile Institute, 1956. 'Handbook of Asbestos Textiles', Asbestos Textile Institute, Philadelphia, 2nd edition, 1961.

important use in the reinforcement of polymers as well as being the basis for many inorganic textiles.

Chrysotile can be carded and spun in the pure form, but for commercial conventional yarns other fibres are often blended with it, and many traditional asbestos textiles will be found to contain a proportion of one or more organic fibres. More recently, a pure chrysotile textile, which is smoother and stronger than the conventional type, has been developed from yarn produced via the coagulation of aqueous dispersions of the fibre.

Up to 500 °C, chrysotile is virtually unaffected by heat, but at higher temperatures structural water is gradually lost and the fibre becomes weak and brittle. Dehydroxylation is complete at temperatures near to 800 °C, and a friable mixture of forsterite, Mg_2SiO_4, and silica remains. At still higher temperatures (1100 °C) some enstatite, $MgSiO_3$, is formed. These changes can be represented as follows:

$$2Mg_3Si_2O_5(OH)_4^- \longrightarrow 3Mg_2SiO_4^- + SiO_2^- + 4H_2O$$

$$3Mg_2SiO_4^- + SiO_2 \longrightarrow 2Mg_2SiO_4 + 2MgSiO_3^-.$$

Two members of the amphibole group of minerals may be encountered. Amosite is the fibrous form of the mineral grunerite, a hydrated iron magnesium silicate, and is a constituent of asbestos cement products. Crocidolite, the fibrous form of the alkali amphibole riebeckite, is characterized by its blue colour and resistance to attack by mineral acids; it is no longer imported into the United Kingdom.

All types of asbestos resist prolonged attack by alkali, but chrysotile, in contrast to crocidolite, is attacked fairly readily even by dilute acids which progressively remove magnesium atoms and eventually leave a silicaceous residue.

The characteristics of the three major types of asbestos are summarized in Table 2.1 below.

TABLE 2.1
Properties of Asbestos Fibres

	Chrysotile (white)	Crocidolite (blue)	Amosite (brown)
Basic composition	Hydrated magnesium silicate	Hydrated silicate of iron and sodium	Hydrated silicate of iron and magnesium
Approximate formula	$Mg_3(Si_2O_5)(OH)_4$	$Na_2Fe_3^{2+}Fe_2^{3+}$ $Si_8O_{22}(OH, F)_2$	$(Fe^{2+})_4(Fe^{2+}, Mg)_3$ $Si_8O_{22}(OH)_2$
Colour of crude rock	Deep green	Blue	Mid-brown
Texture of fibre	Silky, soft	Harsh	Coarse
Flexibility and spinning properties	Excellent	Fair	Poor
Major properties	Flexible, heat-resistant, stiff, strong, alkali-resistant	Flexible, heat-resistant, stiff, strong, acid-resistant	Brittle, long fibres, acid-resistant

2.2 Man-made Fibres

The introduction of so many modified fibres has made the task of identifying man-made fibres more difficult in recent years. While microscopical examination, especially when using some of the more sophisticated techniques, is a useful method it is still advisable to use some of the wide range of chemical and physical tests to supplement the optical ones in order to obtain a reliable diagnosis.

Chemical Properties

The chemical properties of man-made fibres are important aids to identification with so many synthetic-polymer fibres on the market. Of the chemical tests available, the burning test is still of great value as it gives much useful information. The fibres should be advanced slowly towards a small, carefully controlled flame and finally into it. During this operation the behaviour of the fibres should be accurately noted and recorded; whether they melt and form a bead that shrinks from the flame, or can be pushed into the flame, and whether, if they burn, beads are formed on the ends of the burnt fibres. Some fibres, such as nylon, shrink from the flame only if a fine strand is used. Others, for example, the regenerated cellulose fibres, will not melt under any circumstances, and consequently form no bead. Other techniques for this type of examination are (i) to use a small ignition tube or (ii) to heat the fibre on a piece of platinum foil above a small flame. Under these conditions, the characteristic odours evolved from the various fibres can be readily detected. Any ash remaining after ignition can then be subjected to qualitative analysis to decide if a delustring agent is present; the most frequently used being titanium dioxide. Solubility tests can usually be carried out in test-tubes, however, more information may be obtained if the tests are made on a slide, preferably one with a cavity, the effect being observed through a light microscope. Microscopical observation is essential for fibre blends. Characteristic phenomena may be observed, as some fibres disintegrate and dissolve, and some swell and dissolve, whereas others dissolve without appreciable swelling. For example, in cuprammonium hydroxide solution, regenerated viscose and cotton swell and dissolve, but if these fibres have been rendered resistant to creasing by a resin or other finishing treatment they do not generally dissolve. Clear indication of the type of internal structure can be observed during this test under a light microscope, for example, natural cellulose fibres with spiral structure show characteristic twisting prior to dissolution.

Numerous staining tests have been proposed. However, these are valuable only when applied to undyed and untreated fibres, otherwise they are liable to give confusing results.

Physical Properties

Physical examination is a useful tool in the identification of man-made fibres and several techniques are available to the analyst. Observation of properties, such as refractive index, strength and extensibility in the wet and dry states, and anisotropic swelling, can give rapid and accurate information as to the history of fibres. Cross-sections of man-made fibres can no longer be treated as being distinctive in the light of recent developments, but the fibre outline and the presence or absence of heterogeneities in the fibre substance all give important indications as to type of fibre. For example, a heavily serrated periphery indicates that the filaments have been produced from a comparatively dilute solution by rapid coagulation, for example, by the normal viscose and acetate processes. A smooth outline indicates either that the filaments have been produced without lateral contraction (from loss of solvent), as with polyamide spun from a melt, or that the process of coagulation has been slow and uniform as with cuprammonium fibres and some types of modified viscose. Nowadays synthetic fibres are often produced with special cross-sectional shapes, which can vary even within a single generic type or brand, and great care must be exercised when drawing conclusions from fibre cross-sections.

Heterogeneities in the filaments are seen most clearly in thin sections, preferably not more than 10 μm thick. Pigments, delustring agents, particles of agents added to modify dyeing properties, and gas bubbles may be seen in this way. The fineness of filaments is also best observed in cross-section.

Nomenclature of Man-made Fibres

Man-made fibres are marketed under a bewildering number of terms and names. There are the brand names as well as the names by which technologists know the fibre, these latter generally being the chemical or process names. Thus there is a need for a family or generic name to group together all fibres of similar chemical type. These generic names must be capable of strict definition, but yet be reasonably easy to understand by all people who will need to use them.

Generic names have recently been regularized for international use in both ISO and BSI documents[22]. Table 2.2 lists these generic names, together with typical examples of trade names.

TABLE 2.2
Generic Names of Man-made Fibres

Generic Name	Chemical Constitution	Examples of Trade Names
Acetate	Secondary cellulose acetate	Dicel
Acrylic	At least 85% by mass of acrylonitrile	Acrilan, Dralon, Courtelle, Orlon
Alginate	Metallic salts of alginic acid	Calcium alginate
Chlorofibre	At least 50% by mass of poly(vinyl chloride) or poly(vinylidene chloride)	Rhovyl, Leavil, Saran
Cupro	Regenerated cellulose produced by the cuprammonium process	Cuprama
Elastane	At least 85% by mass of polyurethane elastomer	Enkaswing, Lycra, Spanzelle

[22]International Standardization Organization, ISO R/2076. British Standards Institution, BS 4815.

TABLE 2.2 (*continued*)
Generic Names of Man-made Fibres

Generic Name	Chemical Constitution	Examples of Trade Names
Elastodiene	Natural or synthetic polyisoprene	—
Fluorofibre	Polymer made from fluorocarbon monomer	Polifen, Teflon
Glass	Mixed silicates	Fibreglass, Marglass
Modacrylic	Polymer made from 50–85% by mass of acrylonitrile	Dynel, Teklan
Modal	Regenerated cellulose made by a high-wet-modulus process	Vincel
Nylon or polyamide	Polymer with the recurring functional group –CO–NH–	ICI Nylon, Enkalon, Perlon, Celon
Polycarbamide	Polymer with recurring functional group –NH–CO–NH–	—
Polyester	At least 85% by mass of an ester of a diol and terephthalic acid	Terylene, Dacron, Trevira
Polyethylene	Polyethylene polymer	Courlene, Drylene
Polypropylene	Polyethylene where one carbon in two carries a methyl side chain	Fibrite, Meraklon
Polyurethane	Polymer with the recurring functional group –O–CO–NH–	—
Triacetate	Cellulose with at least 92% by mass of hydroxyl groups acetylated	Tricel, Arnel
Trivinyl	Vinyl terpolymer of which no single component forms more than 50% by mass	—
Vinylal	Poly(vinyl alcohol)	Kuralon, Vinylon
Viscose	Regenerated cellulose obtained by the viscose process	Fibro, Sarille, Evlan

2.2.1　Natural-polymer Fibres

2.2.1.1　*Regenerated Protein Fibres*

Casein (e.g., Merinova)

Regenerated protein fibres produced from casein are manufactured by wet extrusion processes in which an alkaline solution of the protein is coagulated by passage through acid liquors.

The fibres receive a 'hardening' treatment with formaldehyde or benzaldehyde and metal salts to reduce swelling in aqueous media and improve their resistance to alkalis. The cross-sections are almost round with a slight indentation. These fibres are employed as staple in blends with other fibres for processing on cotton, worsted, or woollen spinning machinery, and in felts.

Fibres from casein are usually slightly cream in colour and they give staining reactions similar to those for animal fibres, e.g., Millon's reagent (see Section 5.17.10) gives varying shades of brown, the Xanthoproteic reaction gives orange shades (see p. 6), and picric acid stains the fibres yellow.

The fibres are insoluble in cold, concentrated hydrochloric acid, which dissolves *Bombyx* silk. When warmed in this acid, Merinova gives a purple colour, whereas silk gives an almost colourless solution. When hydrolysed with hot, dilute hydrochloric acid, the formaldehyde that is liberated may be detected in the hydrolysate.

Indigo Carmine (Colour Index Acid Blue 74) stains Merinova dark blue, whereas wool is only slightly coloured.

2.2.1.2 Regenerated Cellulose Fibres

Regenerated cellulose fibres are made by a process that involves a number of complex chemical and physical reactions, permitting considerable versatility in both the process and the product. The most important fibres within this group are produced by the viscose process, which involves extruding viscose (an alkaline solution of sodium cellulose xanthate formed by reacting wood pulp with caustic soda and then with carbon disulphide) through the fine holes of a spinneret into an aqueous coagulating bath containing sulphuric acid and salts, commonly sodium and zinc sulphates. Here the alkali in the viscose is neutralized, the filaments coagulate, and the regeneration of pure cellulose is initiated. Cuprammonium yarns are prepared by extruding a solution of cellulose in cuprammonium hydroxide into water. These stages are collectively referred to as the spinning process and are followed by further stages of stretching and fixation, relaxation, washing, finishing, drying, and collection. Various types of regenerated viscose fibres are made, often with quite different mechanical properties and differing among themselves and from cuprammonium and other types of cellulosic fibre in the fine structure of the fibre and the degree of polymerization of the cellulose.

The products of both the cuprammonium and viscose processes have been commonly referred to collectively as 'rayon'. This term is no longer acceptable for labelling goods containing these fibres, which must now be designated 'cupro', 'viscose', or 'modal' (see Table 2.2).

Viscose and Modal Fibres (e.g., Avril, Darelle, Evlan, Fibro, Sarille, Tenasco, Vincel, Zantrel) (Figs 89–100, 166)

Regenerated viscose is produced in a wide range of linear densities either as continuous-filament yarn or in tow or staple forms; the staple form may be crimped and is available in various staple lengths. All types may be bright or mass-pigmented to produce either delustred fibres by adding titanium dioxide particles to the spinning solution or coloured fibres by adding coloured pigments. Until the 1930s all the viscose rayon produced was essentially of one type, though considerable variations between samples of different origin did occur as a result of different production techniques. At the present time this type, referred to as 'viscose' or regular rayon, either filament or staple, still forms an important part of the total production, but industrial yarns and staple fibres developed for specific end uses are now increasingly used.

During the late 1930s yarn of higher tenacity was introduced for use in tyre cords and industrial products and progressive increases in tenacity have continued to be made since then. Changes in spinning conditions to give the fibre asymmetrical properties across its cross-section have resulted in crimped viscose, produced in the form of staple fibre. All these types possess relatively low initial moduli in the wet state, and modifications in both the viscose preparation and the coagulation and stretching sequences have enabled 'modal' or high-wet-modulus types to be produced, again mainly as staple fibre. Products of even higher wet moduli (the so called 'polynosic' fibres) are also available, but these have not achieved such wide acceptability as the modal types because of other technical limitations.

The most recent development has been the introduction of yarns and staple fibres, which, unlike other cellulosics, possess flame-retardant properties when used in suitable constructions. This is achieved by including additives (usually organic compounds containing phosphorus and halogens) in the viscose dope prior to spinning and all types of viscose fibres can now be produced with flame-retardant properties (F.R.).

In longitudinal view as seen under the microscope, regular viscose, both filament and staple, is fairly uniform in diameter with continuous striations running closely parallel to the fibre edges. Pigment particles, either for delustring or mass-coloration, are visible under moderate magnification, and in F.R. fibres globules of additive can be seen inside the fibre and as protrusions on the fibre surface. The cross-section of regular viscose shows serrations at the periphery of the fibre; fine filaments showing a more uniformly serrated outline than coarse filaments. In the cross-section a skin and a core region can be distinguished by suitable staining or optical methods and it can be shown that dyes have different diffusion and retention behaviour in these regions. Crimped fibres possess this skin–core structure in an asymmetric form, and high-tenacity types, whether continuous-filament or staple, tend to have a predominantly skin structure. These are further differentiated by their smooth-edged, and round or bean-shaped cross-sectional outline, with correspondingly fewer striations in longitudinal view.

Like all forms of cellulose, regenerated viscose is dyed by direct dyes, some of which cause fairly permanent staining at a low temperature. Shirlastain fibre identification stains are frequently useful in distinguishing viscose from cupro (e.g., Shirlastain A), cellulose acetate, and fibres of different chemical constitution. Such stains have also a limited application in distinguishing between different types of regenerated viscose, e.g., Shirlastain F produces brown colorations with both regular viscose and some H.W.M. types, but the latter are redder.

Table VIII (g) lists the main groups of yarns and fibres that are currently produced and gives typical properties and some features useful for differentiation.

Regular viscose rayon, as continuous-filament yarn, has a tenacity measured air-dry (conventionally 65 ± 2% relative humidity and 20 ± 2 °C) of up to 196 mN/tex with a breaking extension of 20–30%. In the wet state the tenacity is only half the air-dry value. Because of the importance of viscose yarns for the reinforcement of tyres and other industrial products, work over the years has concentrated on improving tenacity and optimising modulus. As a result changes in the composition of the spinning bath, stretching conditions, and refinements in spinning techniques have led to the development of tyre yarns with air-dry tenacity of up to 471 mN/tex and wet tenacity of 283 mN/tex. This process ultimately gives filaments with a smooth cross-section and with an 'all-skin' structure as defined by staining tests (see Section 5.6).

In the production of staple fibre, again, by adjusting the chemical composition of the viscose, the spinning bath, and the stretch conditions, various new types of staple fibre are produced having differences in the proportion of the fibre made up by crystalline regions, in the crystallite size, and in the degree of orientation. In addition, each type has a characteristic microscopical appearance. The newer fibres have been developed with a variety of different properties making each fibre more suitable for specific end-uses. Crimped fibres have a softer, more bulky handle, and fibres with a high-wet-modulus have greater fabric stability; fibres with a high work of rupture possess enhanced abrasion resistance. These differences govern not only the mechanical and physical properties of the fibres but they afford a practical means of identification that is made use of in Tables VIII (c)–VIII (g).

The dyeability of regenerated viscose can be enhanced to some extent by modifying the chemical treatment of the fibre during manufacture, but this is done more effectively by incorporating a special additive in the spinning dope. When a blend of deep-dyeing and normal viscose in yarn or fabric form, is treated with a single dye, two distinct tones are obtained. The reactivity of the deep-dyeing fibre to dyes which are normally used to dye wool has led to the use of the term 'wool-dyeing' to describe this type of fibre and this characteristic can be used to differentiate deep-dyeing fibre from normal regenerated viscose (see Table VIII (f)).

Most domestic and apparel fabrics containing staple viscose rayon fibres and some made from continuous-filament yarns receive a resin finishing treatment to improve resistance to creasing during wear and in washing and to reduce shrinkage on wetting. Resins are also used to fix surface effects (schreiner, chintz, and embossed finishes), or achieve shape retention in garments by durable-press treatments.

Urea–formaldehyde resins are still most commonly used for crease-resist finishes on 100% viscose fabrics and work by depositing a high-molecular-weight elastic polymer on the fibres. Because this involves little formation of bonds between the resin and the fibre the finish is not permanently wash fast and at the present time the move is increasingly towards the use of 'reactive' resins, usually based on cyclic ethylene urea (e.g., dihydroxy dimethyl ethylene urea (DHDMEU)) which cross-link the cellulose molecules inside the fibres. Reactive resins are frequently found on fabrics containing blends of viscose with nylon, polyester, or acetate fibres, and DHDMEU, in particular, is used where the blend fabric is given a durable-press finish.

These finishes are not usually visible as a coating on the fibre, and do not modify the fibre appearance under the microscope. They reduce aqueous swelling, prevent dissolution in cuprammonium hydroxide, and also modify the colour produced by mixed-dye stains. Their presence can often be detected by the slight fishy odour of the burning material. More reliably, the nitrogen of the amino compound can be detected by the soda-lime or sodium fusion test, and the formaldehyde, after acid hydrolysis, by the chromotropic acid test. Most resin finishes can largely be removed by boiling in dilute mineral acids. Positive identification of the various types of resin is difficult but some success has been achieved using chromatographic analysis (see Section 5.3).

c

Specialized finishes, designed, for example, to confer oil or water repellent properties on fabrics for specific end-uses are also often applied at the resin finishing stage. Among the most important of these are the various treatments for rendering cellulosic fabrics flame retardant, among the most common of which are treatments based on phosphonium chloride and phosphonopropionamide derivatives. In general, the F.R. properties of fabrics finished in this way are not as durable to repeated laundering as fabrics made from fibres which are themselves inherently flame retardant.

Cuprammonium Fibres (Figs 101, 102)

Cuprammonium yarns are manufactured by extruding a solution of cellulose (either from purified wood pulp or cotton linters) in aqueous cuprammonium hydroxide into water and then treating with acid. The cellulose is precipitated in the form of cylindrical filaments that are subjected to high stretch during the coagulation process. The filaments are then treated with dilute sulphuric acid to remove copper salts, washed with water, and finally treated with lubricating or softening agents before drying.

During the coagulation process, the filaments make contact with each other while still plastic and this produces cross-sections that are slightly polygonal, and some filament adhesion that is, however, broken during subsequent mechanical treatments.

The structure of the fibre is more open than that of viscose rayon, i.e., the pore size is larger so that direct dyes diffuse more rapidly into cuprammonium than into viscose. This difference forms the basis of numerous staining tests which have been employed to distinguish between the two fibres (e.g., Shirlastain A).

The fibres may be delustred by the addition of pigments to the cuprammonium spinning solution or by after-treating the fibres or yarns.

The cuprammonium process has become progressively less economical in recent years and cupro yarns are now encountered only rarely.

2.2.1.3 Cellulose Acetate Fibres

Cellulose diacetate (e.g., Dicel, Arnel, Silene) and Cellulose triacetate (e.g., Tricel, Arnel, Rhonel) (Figs 103–106, 167)

Two principal varieties of acetate fibres are now manufactured by dry-spinning techniques. Both types can be produced in the bright form and also delustred by the incorporation of a pigment. The triacetate fibre (tertiary cellulose acetate; acetic acid yield 61·5–62%) is produced from a solution of cellulose triacetate in methylene chloride, whereas the fibre with an acetic acid yield of 54–55% is spun from an acetone solution of a secondary, partially hydrolysed, acetate. The fibres are smooth and in the absence of delustrant, transparent. Cross-sections usually have a clover-leaf shape with large lobes, although peanut-shaped sections also occur. Longitudinally, the fibres appear smooth with a few striations parallel to the fibre axis. The birefringence of cellulose acetate fibres is very low, and thus it is possible to find a single liquid with a refractive index close to that of the fibres in which they are very nearly invisible, e.g., liquid paraffin (refractive index 1·47).

The differences in degree of acetylation affect the chemical properties of the fibres, particularly heat stability and solubility. In contrast to unmodified cellulose, acetate fibres are thermoplastic and use is made of this property, particularly in the case of triacetate, to heat-set crimp in yarns and durable pleats or creases in fabrics. Shapes introduced into triacetate fibres and fabrics can be set by heating above 170 °C or by pressure steaming.

The diacetate fibres can be delustred by treatment with hot aqueous solutions of, for example, soap, phenol, or cyclohexanol. Soap delustring is now the only process of significance and commercially even this is little used. Triacetate is unaffected by these treatments.

Cellulose acetate fibres have little or no affinity for direct or acid dyes, and are usually dyed with disperse dyes. These characteristic dyeing properties can form the basis of staining tests, e.g., Shirlastain A.

Fibres of cellulose triacetate are sometimes partly saponified with alkali in order to give durable antistatic properties, a different fabric handle, a reduction in gas fume fading with certain dyestuffs, and a greater acceptability to chemical finishes. In this case, the saponification affects only the outer cuticle of the fibres (the so-called 'S' finish), the inner parts consisting of unchanged cellulose triacetate. If such fibres are stained with a mixture of a disperse and a direct dye of different colours, a clear line of demarcation between the unsaponified core and the saponified outer cuticle can be seen in their cross-sections. Likewise, in their cross-sections, the marked birefringence of the saponified cellulose layer clearly differentiates it when examined between crossed polars under the microscope. Extraction of the partially saponified fibres with acetone dissolves away the core of cellulose triacetate, leaving hollow tubular fibres.

The triacetate is soluble in methylene chloride, m-cresol, 90% phenol, and 100% acetone, but insoluble in 70% acetone and concentrated hydrochloric acid. Cellulose diacetate is soluble in m-cresol and 90% phenol, and also soluble in 70% acetone and concentrated hydrochloric acid, but is substantially insoluble in methylene chloride.

2.2.1.4 Miscellaneous

Alginate Fibres (Figs 107, 108)

These fibres have serrated outlines with occasional flats where two or more fibres have been pressed together while still plastic during manufacture.

At present, only calcium alginate fibre is commercially available. It is non-inflammable on account of its high metal content. Calcium alginate yarn is marketed chiefly for uses in which its alkali-solubility is turned to practical advantage, e.g., as a linking thread in the hosiery trade or as a scaffolding thread in light-weight fabrics. Nowadays it is mainly produced for surgical end uses (dressings, swabs) where it is usually first converted to the sodium salt by ion-exchange, to give a water-soluble product which is readily absorbed in the blood stream.

It is readily identified by its high ash content and its ready gelatinization, followed by dissolution, in sodium carbonate solution.

Paper Yarns (Figs 109, 110)

These yarns are produced from paper that is specially manufactured from kraft or strong sulphite wood pulp, though it is possible to use rag papers. High tensile strength in the machine direction together with uniformity of structure and thickness are desirable properties. The yarn may be reinforced by a strand of hemp, jute, or even steel wire.

Untwisting of the yarn gives, in most cases, a strip of paper from less than 1 mm to upwards of 50 mm in width. In the case of fine yarns it may be impossible to untwist the yarn and thus their identification as paper may depend on the type of fibre present. Some yarns may consist of two paper singles plied together.

In order to avoid the reduction of strength on wetting that is normally characteristic of paper yarns, it is usual to incorporate some wet-strength resin such as urea– or melamine–formaldehyde, in the paper. The rate of wetting may also be reduced by the addition of wax that may be (i) incorporated in the paper, (ii) applied to the paper strips before or after spinning, or (iii) a combination of both processes. Polishing materials or other special agents may also be used.

Identification of the fibres from which the paper was made follows the method normally used by papermakers [23]. Briefly, it involves disintegration of the paper and examination of the stained pulp under a light microscope. A useful stain for this work is zinc chlor-iodide (Herzberg Stain) (see Section 5.17.16).

2.2.2 Synthetic-polymer Fibres

Synthetic-polymer fibres, which are organic in nature, are produced from a variety of polymeric substances that have been (or can be) formed entirely by chemical synthesis. They therefore differ from other man-made fibres, such as those derived from the regeneration or recovery of a naturally occurring polymer, e.g., cellulose or protein, or a modification of the same. It is thus possible to engineer them chemically or physically to suit them particularly for specific end-uses, while the basic properties of natural-polymer fibres can be varied only over relatively narrow limits. Synthetic-polymer fibres presently manufactured include nylons, polyesters, polyolefins, polyurethanes, polyacrylonitriles, and other vinyl polymers and copolymers.

Bicomponent (or conjugate) fibres can also be made from synthetic polymers. In these, each filament contains two different polymers in a definite disposition along its length. These polymers can occur side-by-side, or with one as a core within the other, or one polymer may appear discontinuously as globules or fibrils within a matrix of the other polymer. In these ways the combined properties of both polymers, or the differential in properties between then, can be utilized.

One of the special features in the production of practically all types of organic synthetic-polymer fibres is that, after the initial extrusion processes, they undergo an additional stretching or drawing process for the

[23]C. H. Carpenter, L. Leney, H. A. Core, W. A. Cote Jr, and A. C. Day. 'Papermaking Fibres', State University College of Forestry at Syracuse University, Syracuse, New York 1963.

purpose of greatly increasing the orientation of the molecular structure of the fibre, thus increasing the fibre tenacity.

Although originally many types of synthetic fibres had characteristic cross-sections, various shapes can now be produced at will in the same polymer, and thus cross-sectional shape has ceased to be specific for identifying fibre types. Similarly, although the basic synthetic fibres are normally straight, modifications during manufacture, or afterwards, can be introduced to give 'bulked' filaments, e.g., with a zig-zag or helical crimp, or with some other form of controlled distortion.

Synthetic fibres can have delustring agents added to the polymer prior to extrusion. The amount can be controlled to provide degrees of dullness, for which terms such as semi-dull, dull, and extra dull are common.

The majority of synthetic-fibre materials can be coloured by dyeing or printing techniques. However, with some types of fibre made from hydrophobic polymers, aqueous dyeing in the usual way can be very slow, and a carrier is added to speed up dye penetration to commercially acceptable rates. Alternatively, increase in temperature by dyeing under pressure can in many cases give superior results. With some fibres, e.g., polyester, coloration under hot dry conditions can be used, as in transfer printing. A completely different method is to incorporate insoluble pigments in the melt or solution to be extruded, and high degrees of fastness can be obtained in this way.

Synthetic-polymer fibres are resistant to mildew and bacteriological attack since they cannot provide suitable nutriment. However, sizes, lubricants, and some finishes, even in minute amounts, may permit growth of such biological materials, and while these in general will be removable by washing this does not necessarily apply to any discoloration caused by the organisms. Thus suitable precautions in storage should still be taken with synthetic-fibre materials, e.g., with sailcloths.

Synthetic-polymer fibres, together with other man-made and natural fibres, can be arranged in well defined classes (see Table 2.2), and notes on these are now given on the following pages.

2.2.2.1 Acrylics

(e.g., Acrilan, Courtelle, Creslan, Crylor, Dralon, Orlon, Sayelle (Figs 111–121, 165. Infrared spectra p. 193. Pyrograms pp. 199–201)

These are defined as fibres in which the fibre-forming substance is any long-chain synthetic-polymer composed of at least 85% by mass of acrylonitrile units.

A large number of different acrylic fibres are now in use and most fibre-producing countries now manufacture at least one acrylic fibre. The acrylics are differentiated principally by reason of the small amounts of other vinyl monomers that are copolymerized with the acrylonitrile. This is done to alter and improve the plasticity and dyeing properties of the resultant fibre. In general, all acrylics can be dyed with disperse dyes, but the majority now contain a copolymerized monomer containing an acidic or basic group so that they may be readily dyed with basic or acidic dyes, respectively. These differences in dyeing provide useful information for distinguishing the various fibres that are encountered (see Table V (c)). Information on the composition of the fibres and on the conditions under which they are spun is limited, though the majority are spun from a solution of the polymer. Several companies are now producing conjugate fibres with a bilateral structure. These fibres may vary in cross-section from round to acorn-shaped and are made up of two components that exhibit different shrinkage behaviour towards heat and moisture, such that the differential contraction causes a three-dimensional helical crimp to be formed.

The cross-sectional shapes differ considerably between one type of acrylic fibre and another, between different linear densities of the same type, and even within one and the same yarn. Four fairly well defined shapes can, however, be recognized and in this scheme they are named dogbone, round (including nearly round), bean shaped, and lobed. Examples of these shapes are illustrated in Figs 112, 118, 119, 121. The longitudinal appearance is generally smooth and regular with few striations. Most types of acrylic fibres can be obtained in a bright or delustred form, the most usual delustrant being titanium dioxide. A few manufacturers now produce fibres that are coloured by means of pigment additions to the spinning solution, when the pigment particles can often be seen during microscopical examination of the fibres. Densities range from $1\cdot14$–$1\cdot18$ g/cm^3.

Acrylic fibres burn readily with a luminous smoky flame and a characteristic smell. When heated in a test-tube they gradually darken and ultimately become black; discoloration takes place at about 205 °C.

Fibres that have not been stretched during manufacture do not shrink when heated, but stretched fibres shrink when immersed in boiling water. This property is made use of in processing so-called bulked yarns that are spun from a mixture of stretched and unstretched fibres or bicomponent fibres (see Sayelle, Figs 113, 114). When these yarns are heated the contractions of the stretched fibres induce the unstretched fibre to bulk and crimp, and give the characteristic bulky appearance of the yarn.

The fibres show remarkable resistance to the more common organic solvents although they are all soluble in boiling form–dimethylamide and in dimethyl sulphoxide. The rate of dissolution at a particular temperature, or the temperature at which dissolution occurs in a given time in certain solvents, for example, nitromethane, differs for different types and provides useful evidence for identification of different types. Resistance to acids is high but acrylic fibres are dissolved in boiling concentrated sulphuric and nitric acids. They are more readily attacked by alkaline solutions although their resistance at low temperatures is good. A common feature of all acrylic fibres is the high percentage of nitrogen present, associated with the cyanide groups. Nitrogen may be readily detected by the soda-lime test.

Acrylic fibres are normally in either staple or tow form and they are widely used alone or in blends with both natural or man-made fibres. Some continuous-filament yarns are available, e.g., Crylor and Dralon. Because of their extremely versatile nature acrylic fibres may be found in almost any type of textile application.

2.2.2.2   Chlorofibres

Poly(vinyl chloride) (e.g., Fibravyl, Rhovyl, Thermovyl, Leavil), chlorinated poly(vinyl chloride) (e.g., PeCe, Piviacid), copolymers of vinylidene chloride and vinyl chloride (e.g., Saran, Velan) (Figs 122–125, 164)

These fibres are made from poly(vinyl chloride), poly(vinyl chloride) after-chlorinated to increase its chlorine content, and poly(vinylidene chloride), and they contain from 53–70% of chlorine. They are made as multifilament yarns, monofilament, and staple, and in types with different capacities for shrinkage when heated. All have a high resistance to cold water, acids, and alkalis, but they are swollen by, or dissolve in, many chlorinated hydrocarbons and aromatic solvents. They soften and shrink at fairly low temperatures, but do not burn. Their resistance to light is good and their main uses are where their inertness towards water and light, their thermal shrinkage, and their non-inflammability can be exploited. The fibres can be dyed but for many purposes they are mass-pigmented.

Poly(vinyl chloride) in the form of Rhovyl (filament) and Fibravyl (staple) contains about 54% of chlorine and has a density of 1·38–1·40 g/cm³. It shrinks by 40% or more at 100 °C and melts at about 160 °C. The fibres may be either melt-spun or dry-spun from mixtures of acetone and carbon bisulphide. The high thermal shrinkage is utilized to produce bulky and stretchable articles in blends with wool and other fibres.

Fibres are also produced in pre-shrunk form, e.g., Thermovyl, and may have tenacities below 88 mN/tex and extensibilities up to nearly 200%. Because the fibres in this thermally-shrunk state are disoriented their density is 1·38 g/cm³, lower than that of unshrunk fibres.

A fibre (Leavil) with improved heat stability and improved resistance to chlorinated solvents has been produced by a low-temperature polymerization process to give syndiotactic poly(vinyl chloride).

Chlorinated poly(vinyl chloride) fibres have an intermediate chlorine content and their properties are broadly similar to the others in the group.

Fibres from poly(vinylidene chloride) usually contain some vinyl chloride (not more than 15%) as a copolymer, and perhaps a little acrylonitrile. The chlorine content is about 63%. The fibres are melt-spun. Their density is about 1·7 g/cm³ and melting point about 160 °C.

In practice the fibres are recognized by the high content of chlorine, the virtual absence of nitrogen, and their low melting points. They are differentiated by density, to some extent by melting point, by their solubility behaviour in particular solvents, and by chlorine content. They stain feebly in the ordinary fibre stains, which are of little value in this connexion, though a staining test with morpholine is useful for distinguishing poly(vinylidene chloride) from the rest. Other chlorine-containing fibres are Vinyon HH and Kohjin Cordelan. Vinyon HH is made from a copolymer of vinyl chloride and vinyl acetate that is solvent-spun; the fibres dissolve in acetone, and were formerly of peanut-shaped cross-section but later types are round. The fibre is used only in small quantity, e.g., for purposes where chemical resistance and its capacity for shrinkage and heat-bonding are important; Kohjin Cordelan is emulsion-spun from poly(vinyl alcohol) and poly(vinyl chloride), and differs from normal poly(vinyl alcohol) in possessing lower moisture regain (3%) and higher density (1·32 g/cm³) owing to its vinyl chloride content.

2.2.2.3   Fluorofibres

Polytetrafluoroethylene (e.g., Teflon TFE), fluorinated ethylene propylene co-polymer (e.g., Teflon FEP) (Fig. 126)

Polytetrafluoroethylene is obtained by polymerization under pressure of tetrafluoroethylene, a gas liquifying at -77 °C. It is prepared by means of a sintering process and is supplied as filament, tow, or staple. The polymer is a remarkably stable substance, being very resistant to heat radiation, weather, and sunlight. It is insoluble except in certain perfluorinated substances above 300 °C, and is unaffected by water, acids, alkalis, and oxidizing agents, even at high concentrations and high temperatures. It does not burn, and is stable up to 215 °C. At higher temperatures it does not melt but decomposes slowly. The vapours evolved during pyrolysis are strongly acidic and contain hydrogen fluoride, which is highly toxic.

The fibres are usually brown but they can be bleached in boiling nitric acid. Producer bleached fibre is also available. Fluorofibres do not stain in any of the usual fibre stains. Their density is 2·3 g/cm^3, which is higher than that of any other textile fibre of organic origin. The friction against metal is low; the coefficient of friction being approximately one sixth that of nylon. The fibres do not absorb water, and their tensile properties are unaffected by moisture. The tenacity of filament yarn is about 137 mN/tex at room temperature, and falls to about 98 mN/tex at 310 °C.

A copolymer of tetrafluoroethylene and hexafluoropropylene is available as Teflon FEP monofilament in diameters of 120 μm and greater. The properties are similar to those of polytetrafluoroethylene except that FEP is fully thermoplastic with a melting point of 285 °C.

2.2.2.4   Modacrylics

(e.g., Crylor, Dynel, Kanecaron, SEF Modacrylic, Teklan, Verel) (Figs 127–130, 163)

Polyacrylonitrile fibres that contain less than 85%, but more than 35%, of acrylonitrile are usually termed modacrylic fibres. Normally they are copolymers containing chlorine. They are all soluble in butyrolactone at room temperature[24] and some in acetone with warming, and this, together with their chlorine content, distinguishes them from normal acrylic fibres. The chlorine confers a high degree of non-inflammability that is completely fast to washing and does not affect the handle. The fibres may be used in either 100% form or in blend with other fibres where low percentages of the modacrylic fibre give considerable flame resistance to the blend as a whole. Dynel and Kanecaron are similar and contain vinyl chloride as a comonomer, whereas Teklan and probably Verel contain vinylidene chloride. The members of these modacrylic groups can be distinguished by solubility tests, and by their infrared spectra, which are highly characteristic.

---

[24] D. J. Bringardner and P. P. Pritulsky. *Text. World*, 1961, **111**, No. 12, 47.

2.2.2.5 Nylons and Aramids

Nylon 6 (e.g., Celon, Dederon, Enkalon, Perlon), nylon 6.6 (e.g., Blue C, ICI-nylon, Perlon T, Ultron), nylon 6.10 (e.g., Decalon, Perfilon, Riplon, Tecron), nylon 7 (e.g., Enant, Onanth), nylon 11 (e.g., Rilsan, Undekalon), bicomponent nylon (e.g., Cantrece) (Figs 131–139)

Nylons are manufactured fibres consisting of polyamides produced by the condensation of molecules having two amide-forming functional groups. The starting materials consist of two types (i) a mixture of a dicarboxylic acid and diamine and (ii) an ω-amino acid. Nylons produced from the former are referred to by two numerals (e.g., nylon 6.6), the first numeral indicating the number of carbon atoms in the diamine and the second the number of carbon atoms in the carboxylic acid. Nylons produced from an amino acid are referred to by a single numeral (e.g., nylon 6) which indicates the number of carbon atoms in the amino acid. Qiana, a nylon for apparel use, is manufactured from a diamine and dicarboxylic acid, the latter being the 12 acid (dodecanedioic acid). Nylons 6 and 6.6 together account for the great majority of the world's nylon production.

In addition to the simple, or homopolymers, mentioned above, copolymers are also produced. These are composed of two or more structural units that may occur in random sequence in the polymer molecule. A typical starting material would be a mixture of an ω-amino acid, a diamine, and a dicarboxylic acid. A similar nomenclature to that used to describe the homopolymers is used for copolymers, e.g., nylon 6.6 : 6 would indicate that the monomers were a diamine, a dicarboxylic acid, and an amino acid, each containing six carbon atoms. As might be expected from the greater irregularity of their molecular chains, copolymers are generally less crystalline and have a lower melting point than their related homopolymers.

Nylons are usually produced by a melt-spinning process. In addition to monofilament and multifilament yarns, staple fibres, and tows, yarns containing bicomponent fibres are also produced. These generally consist of two polymers lightly fused to form a single yarn. In this arrangement, the polymers generally retain their chemical structure and properties whereas the yarns exhibit physical properties (e.g., bulking and appearance) quite different from those of yarns made from the individual polymers. Bicomponent fibres may be formed from any two suitable polymers and have been made from two nylons or from a nylon and polymer of a different type.

Several bicomponent nylon fibres in staple or continuous-filament form are known. Cantrece is a bicomponent fibre that relies, when heated, on shrinkage differential between nylon 6.6 and a copolyamide to achieve crimp. Bicomponent nylon fibres consisting of a nylon 6.6 core and a nylon 6 sheath are used as the bonding agents in non-woven materials (see Section 2.4).

One type of nylon fibre, used mainly for carpets and tyre-cord yarns, contains fibrils of polyester dispersed in each filament.

After spinning, the yarns are drawn to orient the molecular chains, and yarns of increased strength and elastic recovery and decreased extensibility are thus produced. Various processes are used to improve the bulk and texture of the yarns. Nylon fibres, viewed under a microscope with ordinary illumination, are regular and usually circular in cross-section and appear to be structureless. The opacity of nylon yarns is determined by the amount of titanium dioxide incorporated in the polymer, fully dull yarns containing about 2% titanium dioxide. Originally, nylon yarns were dyeable only with disperse or acid dyestuffs. Dyeing variations, with respect to acid dyestuffs, can be obtained by making various additions to the polymer and similarly nylon fibres can be made dyeable with basic dyestuffs. Unmodified nylon yarns are dyed mainly with acid and pre-metallized dyes. Disperse and reactive dyes are used to a limited extent. Many nylons are now made containing antistatic agents which are generally based on polyethers. The presence of an antistatic agent can be detected using a light microscope and examining a whole mount. The antistatic agent shows as a fine fibrillar structure.

Nylons are resistant to alkalis and insoluble in many solvents. Nylon solvents in common use include phenol, m-cresol, formic acid, and concentrated mineral acids. Acidic solvents hydrolyse nylon, the rate of hydrolysis being dependent on the individual polymer. A useful test for distinguishing between nylons 6 and 6.6 is their solubility in 4·4N hydrochloric acid, nylon 6.6 being insoluble and nylon 6 being soluble.

Distinction between the simple nylons is straightforward and the techniques that may be used include infrared spectroscopy, density differences, and determination of melting points. The latter is also of value in recognizing copolymer yarns or mixed-filament yarns, particularly when the melting characteristics are viewed under a microscope. Many tests that are applicable to the simple nylons can be confusing when applied to mixed-filament yarns or copolymer yarns, and when the second component of a copolymer is a minor one, its presence is frequently undetected.

To ensure complete identification of a nylon fibre it is therefore necessary to effect a complete hydrolysis and to identify all the hydrolysis products. In the case of mixed-filament yarns it is necessary to separate the individual filaments (usually by solubility differences) prior to identification.

Aramid (e.g., Nomex, Conex, Kevlar (originally Fiber B)) (Figs 140–144)

Aramids are aromatic polyamides and the class includes materials in which at least 85% of the amide groups are linked to two aromatic rings.

These are specialist fibres and are noted for their resistance to high temperatures, up to approximately 400 °C.

There are different grades of Kevlar, each having a selected balance of properties (tenacity, extensibility, modulus). This is expressed by a number following the name.

2.2.2.6 Polyesters

Poly(ethylene terephthalate) fibres (e.g., Crimplene, Dacron, Diolen, Fortrel, Grilene, Tergal, Terital, Terylene, Tetoron, Trevira) (Figs 145–147, 162)

Standard Terylene and Dacron were the original members of this group of fibres, all members of which are basically the same chemically, and closely similar in physical properties. In constitution they are condensation polymers of terephthalic acid and ethylene glycol, the particular chemical route taken in polymer manufacture varying with different makers. Fibre production is by melt spinning, and the individual filaments are smooth and, when circular in cross-section, unstriated. Individual fibres and filaments with trilobal or other non-circular cross-section can be produced as required for special effects (e.g., sparkle), and cross-section is thus no positive aid to identification; polyester fibres, being essentially featureless in a visible sense, are not distinguishable in appearance from many other synthetic-polymer fibres. Filament yarn (bulked and unbulked), staple fibre, and tow are produced in a wide range of decitex, lustre, and mass-pigmented colours. Polyester fibres are thermoplastic, and can be heat set to give durable pleats and creases which are resistant to wearing and washing. This property is retained in blended-yarn fabrics, provided there is at least 50% of polyester fibre present. Much of the filament yarn produced is used in a stabilized bulk form (e.g., Crimplene) in knitted goods such as jersey.

Polyester fibre is made in medium tenacity ($<$ 550 mN/tex) and high tenacity ($>$ 550 mN/tex) forms, the latter being obtained originally by alteration in spinning conditions at the expense of extensibility. An alternative method is to use a more highly polymerized polymer, when the extensibility can be retained if desired. Conversely, lower-tenacity staple fibres with a reduced tendency to pill are produced by spinning polymers of lower degrees of polymerization; due to their lower strength such fibres tend to break off rather than form into tenaciously held balls on the fabric surface.

Degree of polymerization can be measured by determining the viscosity of an o-chlorophenol solution of the polyester. This measurement can also be used to detect certain forms of chemical degradation of polyester fibre, i.e., degradation which has resulted from a shortening of the molecular chain without any subsequent uncontrolled re-polymerization. This includes the action of water and steam, acids, and ammoniacal derivatives, but not caustic alkalis (e.g., sodium hydroxide, lime, etc.) or dry heat.

Polyester fibres have high resistance to dry heat, light, oxidizing and reducing agents, acids (including hydrofluoric acid), and micro-organisms. They are, however, significantly attacked by hot concentrated mineral acids and by caustic alkalis. They are insoluble in cold sulphuric acid up to a critical strength of approximately 83%; at acid concentrations above this, polyester fibres dissolve rapidly at room temperature. Unlike nylon, polyester fibre is insoluble in formic acid.

The effect of alkali depends on the type of alkali and the conditions under which it is used. Alkalis based on ammonia and its derivatives degrade the fibre but do not dissolve it, although disintegration will eventually take place. However, other alkalis, e.g., sodium or potassium hydroxide, lime, soda ash, etc., act by dissolving the individual fibres progressively from the surface inwards. If the action is stopped part way, finer-decitex fibres remain. The speed of the reaction depends on the individual-fibre decitex, the individual alkali, its concentration, and the temperature, and with cold dilute solutions attack is only slight. If appreciable attack has taken place, a white precipitate of terephthalic acid is obtained on acidification of the liquor.

Most common organic solvents have little effect, but polyester fibres dissolve at room temperature (or on slight warming) in o-chlorophenol, and also in hot m-cresol. Phenols, as a class, either swell or dissolve polyester fibres.

Polyester fibres are relatively difficult to dye at the boil, and do not in general respond to staining tests. Dyeing can be carried out by using disperse dyes at high temperature (120–130 °C), or at the boil with the assistance of a swelling agent (carrier).

For identification purposes, density (1·36–1·41, mean 1·38 g/cm³) and melting point (approx. 255 °C) are useful features. The most characteristic physical property is however the birefringence (0·160–0·190), which is higher than that of any other normal textile fibre.

A useful confirmatory test for polyester fibre is its resistance to 90% *o*-phosphoric acid. When boiled (215 °C) in this for not more than 1 minute, polyester fibre remains visually unaffected. With few exceptions, notably glass, PTFE, and asbestos, other fibres either dissolve with discoloration, melt, or shrivel to gelatinous brown lumps.

Modified polyester fibres (e.g., Dacron 64, Kodel)

Some types of Dacron contain a small proportion of an acidic component introduced as a co-polymer, and such fibres will, therefore, dye with basic dyes as well as disperse dyes. They can be distinguished from other polyester fibres by suitable staining tests and by their melting points.

A type of Kodel is available in which the diol constituent is 1,4-cyclohexane dimethanol and some of the physical properties of this fibre differ significantly from those of standard poly(ethylene terephthalate). The density (1·22 g/cm³) is lower and the melting point (290–295 °C) is higher. It can also be distinguished from other polyester fibres by its insolubility when boiled for 10 minutes in 10% hydrazine in n-butanol.

Bicomponent fibres can be made in which one or both components may be polyester.

2.2.2.7 Polyolefins

Polyethylene fibres (e.g., Courlene X3, Drylene), polypropylene fibres (e.g., Aberclare, Deltafil, Fibrite, Gymlene, Herculon, Meraklon, Neofil, Polycrest, Pylen Type N15 and Type P-10, Reevon, Spunstron, Tritor)

Polyolefin textiles are based on either polyethylene (polythene) or polypropylene.

High-density polyethylene is produced by a low-pressure process. The polymer density is 0·95 g/cm³ and the melting point 133 °C. The catalyst systems used in polymerization are of a type developed by Ziegler and the polymer is sometimes termed Ziegler polyethylene. Chain branching is much reduced, crystallinity is higher, and fibre tenacities are commonly around 392 mN/tex with low breaking extensions of around 10%. High-density polyethylene is produced mainly in a monofilament form for cordage, ropes, and canvas substitutes.

Polypropylene has a density of 0·90 g/cm³ and a melting range of 160–165 °C. Fibre tenacities are in the range 392–687 mN/tex with breaking extensions usually below 20%. Continuous-filament, staple-fibre, and monofilament types are produced by both melt spinning and film-fibre techniques. Polypropylene fibres with novel cross-sectional design and bicomponent polypropylene textile fibres are known.

Unmodified polypropylene can be dyed with selected disperse colours but their fastness properties do not satisfy the stringent requirements for use in textiles.

Polypropylene polymers modified for dyeing are of three main types.

(*a*) Disperse dyeable, e.g., Fibrite DD, Herculon 2. These fibres are modified by incorporating polymers or polymer-forming substances, at the melt stage, before spinning. The additives that have an affinity for disperse dyes may be of the vinyl pyridine type, polyester, or some other different class.

(*b*) Metal-modified polypropylene fibres that contain various metals have been produced, but the fibre most likely to be encountered contains nickel as part of an organic complex (e.g., Herculon Type 40). This complex acts as a light stabilizer and the nickel acts as a mordant for selected metallizable dyes, e.g., Polypropylene M dyes (Allied Chemical Corporation).

(*c*) Acid dyeable, e.g., Meraklon DL and DO.
The polymer is again modified by incorporating suitable additives at the melt-spinning stage. Both fibres have good affinity for anionic (acid) dyes in general and can be dyed satisfactorily with acid and pre-metallized dyes. The use of certain specified dyebath auxiliaries is recommended in order to assist penetration and improve levelling properties and light fastness.

Polyethylene and polypropylene can be differentiated by their infrared spectra, X-ray diagrams (orientation differences show between low- and high-density polyethylene), and by flotation (in isophorone, of density 0.92 g/cm^3, polyethylene sinks but polypropylene floats). Both polymers are extremely hydrophobic and normally difficult to dye; coloration is achieved mainly by mass-pigmentation.

Conventionally melt-extruded polyolefin fibres are smooth and cylindrical, but fibres or tapes used as substitute yarns can also be prepared from sheet film. Polyolefin textiles from film comprise chiefly:

(*a*) Tapes of regular width (of rectangular cross-sectional shape around 3 mm $\times$ 0.025 mm) extruded as separate tapes from a multi-orifice die; used principally for weaving.

(*b*) Wider tapes (of width 20–50 mm) mainly of polypropylene, capable of splitting longitudinally in a twisting process; used principally for cordage and heavy fabrics.

(*c*) Fine fibre, mainly of polypropylene, produced when a flat tube is mechanically fibrillated to produce a continuous network. This may be used as a yarn or the fibres may be substantially separated by stretch-breaking or cutting into staple fibre. These products are characterized by fibre-branching not found in spinneret-extruded fibre; used mainly as carpet pile.

(*d*) Complex geometric networks forming a non-woven fabric, made chiefly from polyethylene; used mainly for decorative netting, porous bandages, and as a fusible layer in fabric bonding.

(*e*) Continuous filament formed when a striated sheet (produced by embossing, forging, casting, or profile extrusion) is stretched. The fibre is used as continuous filament or cut staple for cordage, or for woven applications where high strength is required.

2.2.2.8 *Polyurethane Elastomers and Rubber*

(e.g., Dorlaston, Enkaswing, Glospan, Lustreen, Lycra, Sarlane, Spanzelle) (Figs 148–152)

Polyurethane and rubber elastomeric fibres exhibit high elongation, about 400–800%, and high recovery and recovery rate. The polymer structure of polyurethane fibres is extremely complex, and involves a block copolymer built up from a urethane-linked and isocyanate-ended polyester or polyether, further reacted, for example, with a diamine, to give a network of extended molecular chains. The fibres are therefore not polymerized urethane as their common terminology suggests.

Polyurethane elastomeric fibres are also referred to by the generic name elastane, which defines fibres composed of at least 85% of a 'segmented' polyurethane. The elastane polymer is made by chain-extending a hydroxy-terminated polyether or polyester that has been isocyanate tipped with a di-isocyanate, either by a diol or a diamine, to give a segmented polymer. One part of the polymer, known as the 'soft' segment consists of polyester or polyether chains that give mobility for large deformations, and the other part, known as the 'hard' segment, consists of urethane or urea units, which give intermolecular bonding to prevent a net molecular flow under stress. Polyester based yarns can be identified by their solubility in boiling, dilute caustic soda solution; polyether based yarns are insoluble.

The polymer is usually solution-spun into multifilament or monofilament yarn. Suitable polymers can be melt-spun. Multifilament yarns show individual filaments that are fused together. Polyurethane elastomers are so similar in extension and elastic properties to rubber that they could not easily be confused with any other fibres, and identification problems are confined to these two materials. Urethane fibres burn with a luminous flame with little smoke, but rubber has a very sooty flame; the characteristic smell of burning rubber is an important distinction. Both materials give a positive test for nitrogen.

Rubber is never encountered as a multifilament yarn, and samples are either of roughly square cross-section indicating that they have been cut from vulcanized sheet, or round indicating that they have been extruded from latex. Both materials often show heavy pigment content under the microscope, but non-pigmented polyurethane elastomeric yarns are now available for use in hosiery.

As raw fibre, the urethane elastomers are a good white, and they can be dyed satisfactorily. Rubber cords are usually a poor white, with either a yellowish or bluish cast, and they are not commercially dyeable. Staining tests on the materials give rather unreliable results owing to poor penetration and differences in sample thickness.

Despite the initially observed similarity in extension and elastic behaviour of urethane elastomers and rubber, there are important tensile differences; the breaking stress of the urethane fibres is several times greater than that of rubber, and at all levels of extension their stress or holding power is higher. The densities of urethane elastomers may be somewhat higher than that of rubber ($1 \cdot 10$ g/cm^3), and values in the range $1 \cdot 00$–$1 \cdot 32$ g/cm^3 have been quoted. The moisture uptake of both materials is very low, and moisture regains are close to 1%.

Polyurethane elastomer yarns may be used bare or wrapped with a conventional yarn, and their relatively high strength also permits their use in core-spun yarns where a usually fine elastomeric yarn is processed so that a sheath of staple fibre is spun around it. Since all elastomeric fibres occur in continuous form the separation of covered or core-spun yarns is possible for identification purposes. The elasticated articles in which the yarns are encountered are mainly of a contour–stretch character such as swimsuits and support garments. Rubber cords are usually wrapped with a conventional yarn; bare yarns or core-spun yarns are not often encountered.

A new type of elastomeric yarn is being developed involving extrusion of nylon and polyurethane together to give a bicomponent yarn which can be used without covering.

2.2.2.9 Vinylals

(e.g., Cremona, Kuralon, Vinylon) (Figs 153, 154)

Vinylal fibres are defined as those in which the fibre-forming substance is any long-chain synthetic-polymer composed of vinyl alcohol with different levels of acetalization. Production is confined mainly to Japan.

The polymer is prepared by the hydrolysis of poly(vinyl acetate), and fibres are produced by extruding an aqueous solution of poly(vinyl alcohol) into a coagulating bath, for example, of Glauber's salt. The resultant fibres are rendered insoluble and improved in mechanical properties by a drawing and then a heating process, followed by an after-treatment with formaldehyde or other aldehydes; resistance to hot water may be improved by a suitable degree of acetalization (e.g., Kuralon). Several modified spinning and drawing processes have been developed in an effort to reduce the abnormalities of molecular structure in the polymer, and so attain higher degrees of crystallinity. Thus tenacities vary from 177 mN/tex for fibres of poorly-ordered structure to at least 589 mN/tex for fibres of improved stereo-regularity of molecular structure.

Vinylal fibres of low tenacity are characterized by a clearly-defined skin–core structure and a folded, peanut-shaped cross-section. High-tenacity fibres usually appear devoid of skin-effect, with a rounded cross-section showing minor indentations, whereas both types of fibre show longitudinal grooves.

Because vinylal fibres possess OH groups their moisture regain is in excess of 5%, an unusual feature among synthetic-polymer fibres. Their hydrophilic nature is associated with a reduction in strength in hot water. Vinylals can be prepared in water-soluble or partially-soluble form for special end-uses, e.g., papermaking.

Normally, vinylal fibres are resistant to cold, dilute acids and alkalis but are soluble in hot, concentrated acids, hydrogen peroxide, and phenols. Poly(vinyl alcohol) liberated from the fibres by acid hydrolysis gives a characteristic blue colour with iodine.

In Japan, vinylals have been applied to all areas of textile usage, but widespread acceptance of the fibres elsewhere is apparently limited by the rather poor wet properties, and difficulties associated with elastic recovery and creasing behaviour.

A modified poly(vinyl alcohol) fibre prepared from a mixed polymer of vinyl alcohol and vinyl chloride is manufactured as Kohjin Cordelan (see Section 2.2.2.2).

2.2.3 Inorganic Fibres

Glass Fibres[25]

Glass fibres are produced in two forms, continuous filaments and staple fibre. As it is debatable whether materials and products based on discontinuous or staple glass fibres can properly be classed as 'textiles', this section concentrates on continuous filaments which are mainly used for reinforcing plastics and rubber, and for fireproof curtains. The fibres are not necessarily long or continuous when used in such applications, but are manufactured as continuous filaments by a melt extrusion technique. This differentiates them from staple glass fibre which is often referred to as 'glass wool' and which is almost entirely used for heat and sound insulation purposes. Although some aspects of the manufacturing technology of the two fibres are similar, the fiberization process used in staple fibre production is different. The aims of this process are to make short, intertwined, bent lengths of glass fibre, by impinging powerful gas or steam jets onto the molten glass as it is extruded through the orifices of the spinneret. It will be realized, therefore, that the manufacture and end product of the two types of glass fibre are sufficiently different for them to require separate treatment.

Continuous filament is produced by mechanically drawing, at high speed, a stream of molten glass vertically downwards from a special furnace to a suitable package for subsequent processing into roving[26], yarn, cords, mat, or fabrics. The diameters of continuous filaments range from 3–16 μm, and the number of filaments in each strand is commonly 400 or 800, but may be as high as 2000. Continuous filament in the form of yarn, roving, woven roving, chopped strands, and chopped-strand mats is widely used for plastics reinforcement. Continuous filament in the form of woven tape and fabric is also used for electrical and thermal insulation. Curtains made from woven continuous-filament fabrics have also established a not insignificant market.

Over 99% of all continuous glass fibre produced is of E-glass composition. Although E-glass was originally developed for electrical applications, electrical uses of E-glass are today only a small proportion of the total market, as it is an almost universally applicable formulation and has become standard for most uses.

Until recently, significant quantities of fibre were made from soda–lime–silica glass (A-glass) in sheet form; as patents covering E-glass prevented producers from using it, or because a local source of cheap sheet-glass scrap was available. As a proportion of the total world glass fibre production it is now insignificant, although for general purpose composites A-glass fibre reinforcement has been shown to be perfectly adequate and can even be cheaper than E-glass to produce.

E-glass suffers from one particular disadvantage; it is susceptible to attack by dilute mineral acids. For this reason a chemically resistant glass (C-glass) was developed for use in composites which would be in contact with acidic materials.

---

[25] K. L. Loewenstein. 'The Manufacturing Technology of Continuous Glass Fibres', Elsevier Scientific Publishing Co., 1973.

[26] In the glass fibre industry, continuous-filament roving is defined as a parallel assembly of strands wound without twist.

A fourth type of glass which should be mentioned is the high strength glass fibre, known as S-glass, which was developed by Owens Corning for use in such specialized applications as rocket motor cases. This glass is difficult and expensive to make and has a very limited application.

The typical composition of the glasses mentioned is given in Table 2.3.

TABLE 2.3
Chemical Composition of Glasses

Constituent	E-glass	C-glass	A-glass	S-glass
SiO_2	55·2	65·0	72·0	65·0
Al_2O_3	14·8	4·0	2·5	25·0
B_2O_3	7·3	5·0	0·5	—
MgO	3·3	3·0	0·9	10·0
CaO	18·7	14·0	9·0	—
Na_2O	0·3	8·5	12·5	—
K_2O	0·2	—	1·5	—
Fe_2O_3	0·3	0·5	0·5	—
F_2	0·3	—	—	—

The fibres are invariably sized or finished. The basic requirement of a fibre size is that it must provide lubrication and protection of the filaments and strands in order to prevent interfilament abrasion, both during the wet conditions occurring during fibre drawing and under the substantially dry conditions of strand conversion into products, and during the use of these products by fabricators. In most cases the fibre size must be formulated to combine lubrication with handling characteristics of the strand, the behaviour of the strand during incorporation into a polymer, and the physical and chemical properties of the glass-fibre–polymer composite constructed from it.

There are two groups of glass fibre sizes. The first and older group consists of the starch–oil sizes, whose function is simply lubrication during fibre drawing and the textile operations of twisting, doubling, and weaving. These are also used to facilitate the dyeing of fabrics for curtains, and the varnishing of glass fibres used for electrical insulation. The second group of sizes is designed for the reinforcement of thermosetting resins, thermoplastics, and rubbers. Formulations in this group are much more complex and are specifically designed according to the handling properties required, the ease with which the glass-fibre strands can be wetted by liquid resin or thermoplastics, and the efficiency with which stresses applied to the composite are transferred to the fibrous reinforcement under the actual conditions of use. Fabrics may be heated to remove the processing sizes and then finished with a variety of silanes and Werner-type compounds (e.g., methacrylato chromic chloride). Other fabric finishes include silicones and phenolics. Partial removal by heat of the starch–oil sizes gives a 'caramelized' finish that is used in conjunction with elastomers.

Glass fibres, in general, are characterized by their high strength, low extensibility up to breaking-point, very low knot strength, and lack of (isotropic) birefringence.

High-performance Fibres[27]

Interest is increasing in the development of fibres having high temperature-resistance. The interest shown in this class of fibre is associated with the development of reinforced plastics for use under high temperatures, especially in the aerospace industry, space technology, and special applications associated with the chemical industry.

Fused silica is made as continuous filaments and also as 'staple' for insulation purposes. The melting temperature is in the order of 1700 °C. Glass fibre can be leached with acid so that a silica residue is left, resulting in a fibre with a minimum silica content of approximately 96% and up to 3% of chromium oxide. The melting point of these fibres is also about 1700 °C.

Fibres are also available which are manufactured from such raw materials as alumino silicates, alumina silicon carbide, and boron nitride. Silica fibres are expensive and their use is, therefore, limited to applications where their chemical or electrical properties make them particularly suitable.

Chrome-stabilized alumino silicate fibres can be used at temperatures of up to 1500 °C. They are still comparatively new and, therefore, relatively expensive.

Alumino silicates are one of the more commonly used group of ceramic fibres, and there are several such fibres now commercially available. The main chemical difference between members of this group of fibres is the content of titanium dioxide (TiO_2) and boric oxide (B_2O_3). In spite of this most of them have similar melting points in the region of 1760 °C. As a class, they are inert to corrosion or attack by steam, oil, and many solvents, up to quite high temperatures. The density of this class of fibre is in the range of 2·65 (g/cm³), which is comparable with asbestos and glass fibres.

Carbon fibres (sometimes known as graphite fibres) may also be classified under the heading of high-performance fibres. Although expensive, they offer outstanding mechanical and thermal properties, and as the price becomes more competitive with conventional engineering materials, it is expected that carbon fibres will become increasingly cost effective.

The method of production is to convert an organic fibre precursor to carbon fibre by controlled pyrolysis in an inert atmosphere at temperatures up to 2500 °C. Commercially, the most important precursor is polyacrylonitrile (PAN) fibre. The PAN process involves a stage in which the fibre is first heated in air prior to carbonization by pyrolysis. This improves both the yield of the process and the properties of the carbon fibre. Cellulosic fibres can also be used as precursors, but the yield and properties are poor. Recent developments have suggested that pitch-based carbon fibres are a commercial possibility.

By adjusting the maximum carbonization temperature, fibres of varying properties may be produced. Their applications range from reinforcement of golf club shafts to high performance aerospace outlets where maximum strength and stiffness for a given mass is required.

Ceramic fibres [28] can be prepared by melting the raw material, kaolin or china clay, in an electric furnace. The thin molten stream, which is closely

[27] A. H. Frazer. 'High Temperature Resistant Fibers', Polymer Symposia No. 19, Interscience Publishers, 1967.
[28] M. H. Lindsey. 'Ceramic Fibre Technology', Morganite Ceramic Fibres Ltd.

temperature controlled, falls into an air blast which causes the material to form a mass of random fibres which are gathered, compressed, and conveyed through an oven which completes their processing into blanket or strip form.

Another technique for producing ceramic fibres is to use an adaptation of the viscose process. A hybrid inorganic cellulosic starting material is dissolved in alkali and is precipitated in fibre form by extrusion into an acid bath. The resultant hydrated structure contains the inorganic component in an orientated, lattice-type arrangement. The cellulose component is either carbonized or burned off altogether and subsequent high-temperature firing fuses the inorganic residue into a quartz-like, insoluble high-temperature fibre.

Single-crystal refractory fibres can be prepared from a number of different starting materials. These monocrystal structures, known as 'whiskers', are of a diameter close to one micrometre with lengths up to several centimetres. They exhibit extreme properties of strength and moduli owing to their practically flawless crystalline structure. 'Whiskers' can be formed of graphite by a growth process in carbon vapour under high pressure in an electric arc; various types of refractory oxides and metals can be formed similarly. The usefulness of this type of fibre is limited by their size, the very high cost, and the difficulty of devising a production system of normal capacity.

Ultra-fine metallic fibres[29] are prepared from fine wires of stainless steel that may be sheathed in a dissimilar alloy and drawn through a die to achieve fibre finenesses in the range 6–12 μm. Such fibres are available in continuous-filament form or as staple. They are mainly used to dissipate static in textile structures such as carpets.

2.3 Metal-foil and Metal-coated Yarns

(e.g., Lurex) (Figs 155–157)

Metal-foil and metal-coated yarns are characterized by a flat ribbon-shape with knife-slit edges. The main constructions of metallic yarns in order of commercial importance are as follows.

1. Monoply yarns slit from polyester film of 12 or 24 μm thickness, metallized, and coated both sides either with clear or coloured lacquer (Lurex C50 and C100) or with heat and chemical resistant resin–lacquer (Lurex TE50 and TE100) (Fig. 155 (*b*)). Lurex TE50 and TE100 are non-tarnishing and have greatly enhanced resistance to scouring and dyeing treatments without loss of the suppleness, brilliance, and yield characteristics of Lurex C50 and C100 product types.
2. Laminated yarns based on one layer of aluminium foil sandwiched between two layers of 12 μm thick polyester film using clear or coloured adhesives (Lurex MF150) (Figs 155 (*c*), 156). This yarn has higher strength and abrasion resistance than previously mentioned yarns.

[29]H. F. Mark, S. M. Atlas, and E. Cernia, eds. 'Man-made Fibres: Science and Technology', Vol. 3, Interscience Publishers, 1968.

3. Monoply yarns made either from 12 μm thick transparent polyester film (Lurex N50) (Fig. 155 (*a*)), or from 12 μm thick polyester film treated with a surface dispersion to give a rainbow effect (Lurex N50 (Irise)).

4. Lurex yarn types C50, N50 (Transparent and Irise), and TE50 can also be obtained supported with two ends of either 17 dtex or 33 dtex monofil nylon (Fig. 157).

The most commonly used yarn width is 0·37 mm (250 dtex), but widths of 0·80 mm (540 dtex) and 0·25 mm (176 dtex) are also available. Metallic yarns are usually described in terms of the nominal thickness of the composite film(s) and not the overall thickness of the yarns; the thickness of the resin–lacquer coating or adhesive layer is ignored.

These yarns are used for decorative effect in both woven and knitted fabrics, fancy yarns, braids, embroidery, trimmings, ribbons, laces, woven labels, furnishing, upholstery draperies, and household textiles.

2.4 Nonwoven Fabrics and Other Nontraditional Fibre Assemblies

Fabrics made by means other than by conventional knitting or weaving processes are finding increasing application. They are now frequently encountered, particularly in the fields of garment interlining, disposable and protective clothing, filter media, sanitary and medical tissues, furnishings and curtains, floor coverings, backing fabrics, cleaning cloths, abrasive products, synthetic leathers, and civil-engineering fabrics.

Such materials are based on known fibre types, which are assembled into sheet structures and are held together in a variety of ways. These can be:

(*a*) mechanical, e.g., stitching, felting, needle punching, or other means of fibre entanglement;

(*b*) thermal, e.g., introduction of a proportion of fibres of lower melting point or use of bicomponent fibres in which the components have different melting points;

(*c*) the use of solvent to soften one component; or

(*d*) chemical, e.g., use of a separate bonding agent.

Surface effects to improve fabric aesthetics, such as drape, etc., can be obtained by traditional textile finishing methods, e.g., dyeing, printing, and embossing, or by such means as needle punching or fibre entanglement brought about in various physical ways, or by employing discontinuous bonding.

The fibrous web employed can be prepared in various ways, but most commonly by so-called dry-laid, wet-laid, or spunbonding processes. The likely method of web formation employed is often indicated by the staple length of the fibres involved. Wet-laid nonwovens are produced from short-staple fibres by techniques closely related to those used in the paper industry. Dry-laid nonwovens use staple fibres of normal textile lengths, and are usually prepared on conventional textile opening, carding, cross-lapping, or air-laying equipment. Spunbonded fabrics are composed of endless filaments and are produced by an integrated process combining fibre spinning, random web formation, and bonding; they can be produced only by the fibre manufacturer.

Analysis of the fibrous components of a nonwoven fabric may be complicated by the presence of, and interference from, chemical bonding agents. However, where such chemicals are not involved, analysis is carried out by the normal methods, which are therefore employed for mechanically bonded fabrics, e.g., needle-punched felts, traditional wool-containing felts and stitch-bonded fabrics, where fabric strength has been developed by subjecting the web to stitching processes with or without the introduction of sewing threads.

Similarly, the analysis of thermal- or solvent-bonded products is relatively uncomplicated, though the method of bonding may not be immediately apparent. Heat bonding is effected by the presence of lower-melting-point thermoplastic fibres in the web, which is subjected to a temperature sufficiently high to render that component soft and adhesive. Examples are the use of lower-melting-point nylon 6 to bond nylon 6.6 webs, the bonding of drawn polyester by using the lower melting point of undrawn polyester, and the use of similar polypropylene systems. Heat bonding, however, is not confined to the bonding of chemically similar fibres, where the presence of the grossly distorted bonding fibres usually identifies the process. Increasingly important are heat-bonded products employing core–sheath bicomponent fibres or filaments, with the lower melting point of the sheath polymer being operative in bonding the fibres; spun-bonded materials can be distinguished from staple materials by their endless filaments and random structure. Heat bonding can be carried out by the use of heat alone, or by a combination of heat and pressure. For example, bicomponent fibres can be bonded by hot embossing to give discrete bonded areas or points, and by the use of suitable designs such textiles can be given a remarkably woven-like appearance.

Fibrous structures can also be bonded using solvent methods, the webs being treated with a solvent or plasticizer which renders at least one of the fibrous components adhesive at room or elevated temperature, the solvent being subsequently removed. Such techniques are used in the production of synthetic-fibre waddings and in bonding certain spunbonded nylon products. Cotton nonwovens can be produced by treatment of the fibre web with caustic soda.

Dry-laid and wet-laid nonwovens are, however, most commonly bonded with added chemical bonding agents. One method of application is by complete saturation of the fibrous web with, e.g., a latex, which results in a microscopically thin film of binder over the surfaces of the individual fibres with a build-up at the fibre cross-over points. Alternatively, the binder can be applied by spraying or by printing in discrete patterned zones, which will give a loftier product with better draping properties.

Since in many cases it is virtually impossible to remove the chemical bonding agent by chemical means without destroying the base fibre, analysis in most cases will require fibres to be removed by physical means, sometimes after soaking in a solvent which at least partially softens the binder. When examining the removed fibres, the possibility of interference from residual binder should constantly be borne in mind. Where the bonding system is sufficiently resistant to solvents it may be possible to obtain a rough estimate of the quantities of fibres present by successive removal of the identified fibres by appropriate solvents, terminating with the binder residue itself.

The binders most commonly encountered in nonwoven fabrics are, in the case of interlinings, acrylates and nitrile and styrene rubbers, which may be modified by inclusion of melamine or urea–formaldehyde resins. These binders, and occasionally those based on poly(vinyl acetate), are also employed in wiping cloths, and sanitary and medical applications. Backing materials may contain these, or possibly natural rubber. Synthetic leathers are almost invariably bonded with nitrile rubber or polyurethane resin, and abrasive products normally with phenol–formaldehyde, melamine–formaldehyde, urea–formaldehyde, thermosetting epoxy resins, or, sometimes, with hard acrylate or vinyl chloride polymers. In flame-resistant materials,

TABLE 2.4
Summary of Nonwoven Fabric Composition

End-use	Fibres present	Method of bonding
Bonded-staple fabrics Felt substitutes	Various	Mechanical interlocking
Coating bases	Various	Stitch bonded
Floorcoverings	Polyester and polyamide or co-polyamide bicomponent fibres	Thermal using differential melting points
	Various	Chemical or thermal using a proportion of core–sheath bicomponent fibres
Disposable cleaning cloths	Various	Discontinuous chemical bonding
Interlinings	Viscose, nylon and viscose, polyester and viscose	Chemical bonding
Filter cloths	Viscose	Chemical bonding
Leather substitutes	Various	Chemical bonding
Spunbonded fabrics Interlinings	Polyester and co-polyester	Thermal using differential melting points
Primary tufting base	Polypropylene with drawn and undrawn segments	Thermal using differential melting points of the segments
	Nylon–polyester bicomponent filaments	Thermal using differential sheath–core melting points
Domestic textiles (upholstery, table cloths, etc.)	Nylon 6–nylon 6.6 bicomponent filaments	Thermal using differential sheath–core melting points
Civil engineering fabrics	Polypropylene and polypropylene–nylon 6.6 bicomponent filaments	Thermal using differential sheath–core melting points
	Polyester or polypropylene	Mechanical interlocking by needle punching

poly(vinyl chloride), poly(vinylidene chloride), and their copolymers, as well as chloroprene rubbers, may be encountered. Viscose materials bonded with cellulose xanthate are much used for liquid filtration.

Garment interlinings—woven, knitted, or nonwoven—are often in the form of 'fusible' interlinings, carrying a thermoplastic adhesive on one or both surfaces. This adhesive may be present as random particles or in various patterns, particularly dots, or as an adhesive web or continuous coating. By far the most commonly used adhesives in such applications are polyamides, poly(vinyl chloride), polyethylene, and, to a lesser extent, poly(vinyl acetate) and cellulose acetate.

Table 2.4 gives some typical examples of non-traditional fibre assemblies, but it is not limiting in any way.

3 Photomicrographs

The majority of the photomicrographs have been prepared using the light microscope; those prepared using the electron microscope are identified as follows:

    TEM—Transmission electron microscope
    SEM—Scanning electron microscope.

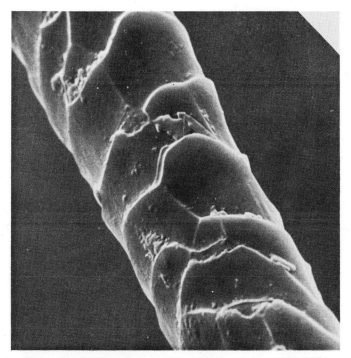

Fig. 1. Scottish Blackface wool fibre. Whole mount × 1300. SEM.

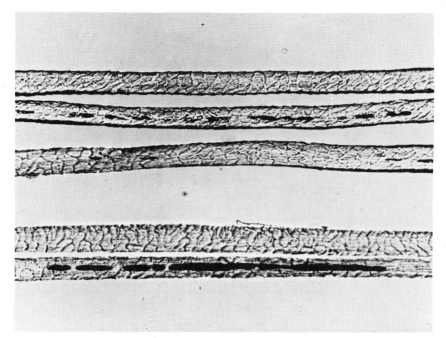

Fig. 2. Wool fibres from 46s top. Whole mount × 180. Note the presence of fragmental medullae.

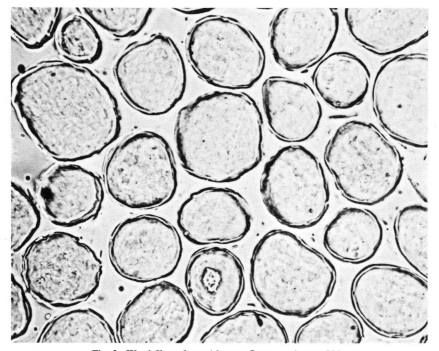

Fig. 3. Wool fibres from 46s top. Cross-section × 500.

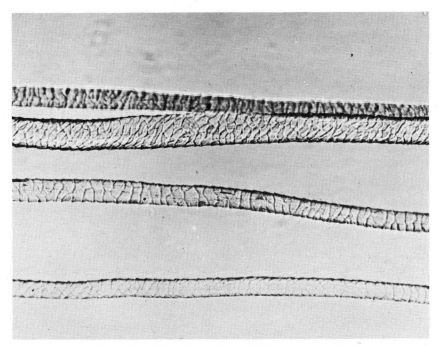

Fig. 4. Wool fibres from 64s top. Whole mount × 180.

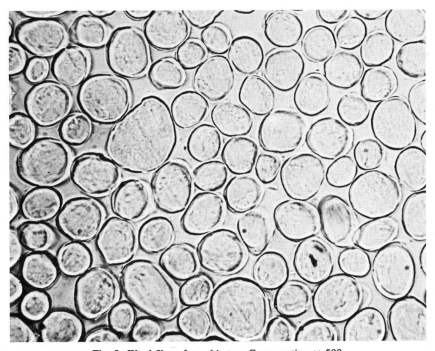

Fig. 5. Wool fibres from 64s top. Cross-section × 500.

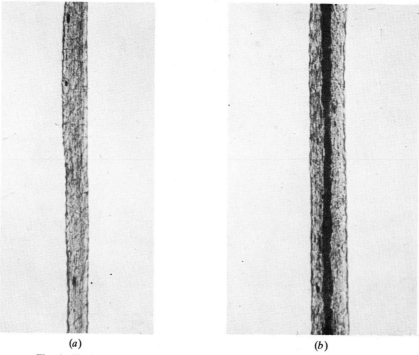

(*a*)   (*b*)

Fig. 6. Shetland wool fibres. Whole mounts × 180.
(*a*) Fine fibre—sparsely pigmented.
(*b*) Coarse fibre—sparsely pigmented, with continuous medulla.

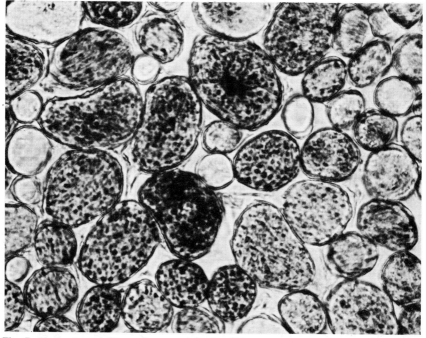

Fig. 7. Shetland wool fibres. Cross-section × 500. Some fibres are pigmented and some are medullated.

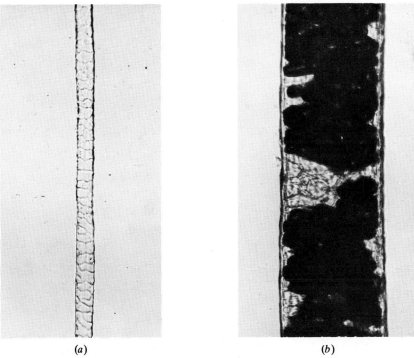

(a) (b)

Fig. 8. Scottish Blackface wool fibres. Whole mount × 180.
    (a) Fine fibre.
    (b) Coarse fibre—medulla is partly air-filled, and partly in-filled with mountant.

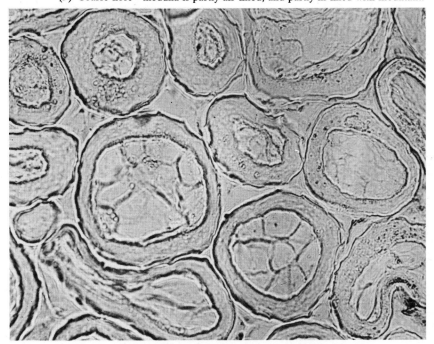

Fig. 9. Scottish Blackface wool fibres. Cross-section × 500.

Fig. 10. Root and tip of kemp fibre. Whole mount × 180.

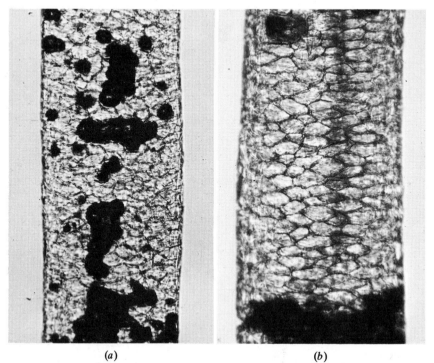

(a)   (b)

Fig. 11. Wool kemp fibres. Whole mount × 180.
(a) Mounted in liquid paraffin. (b) Mounted in lacto–picro–phenol.

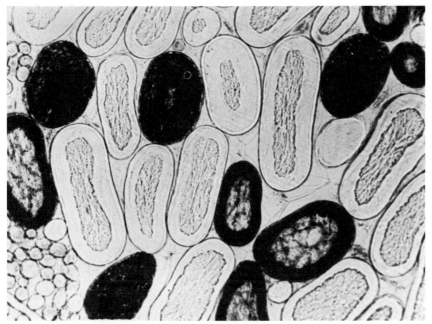

Fig. 12. Fibres from leg of Swaledale sheep. Cross-section × 250.

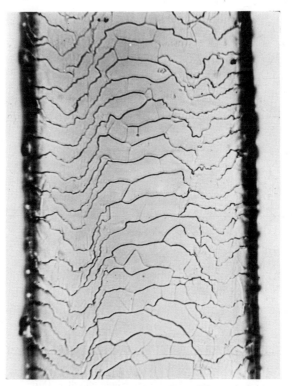

Fig. 13. Swaledale sheep, axillary region. Cast of the scale pattern × 500.

Fig. 14. Skin wool. Root-ends of fibres from the sweating process × 75.

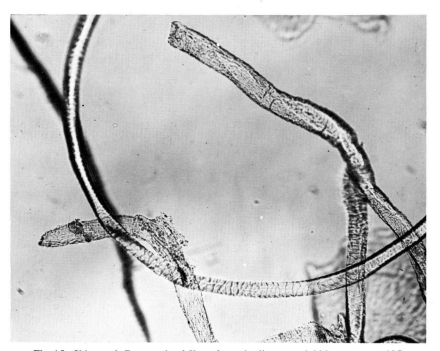

Fig. 15. Skin wool. Root-ends of fibres from the lime or sulphide process × 125.

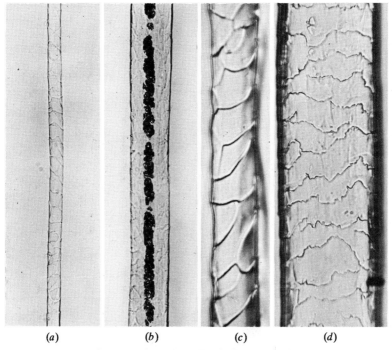

(a) (b) (c) (d)

Fig. 16. Mohair fibres. Whole mount × 180.
(a) Fine fibre. (b) Coarse fibre.
Casts of the scale patterns × 500.
(c) Fine fibre. (d) Coarse fibre.

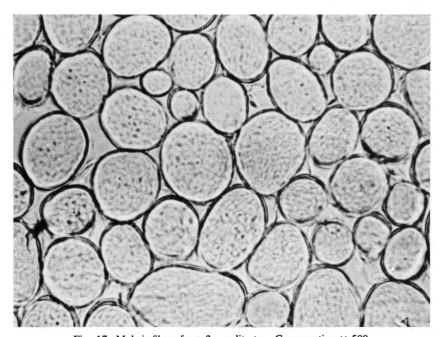

Fig. 17. Mohair fibres from 2s quality top. Cross-section × 500.

E

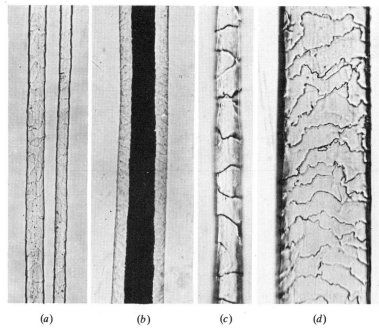

(a)   (b)   (c)   (d)

Fig. 18. Cashmere fibres. Whole mounts × 180.
(a) Fine fibre. (b) Coarse fibre.
Casts of the scale patterns × 500.
(c) Fine fibre. (d) Coarse fibre.

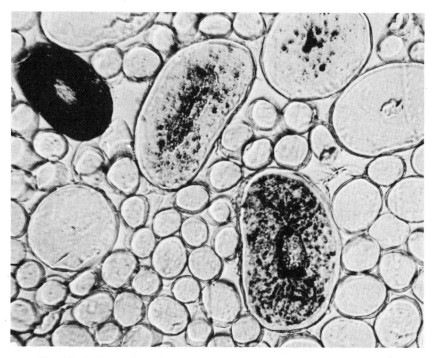

Fig. 19. Cashmere fibres from Persian cashmere fleece. Cross-section × 500.

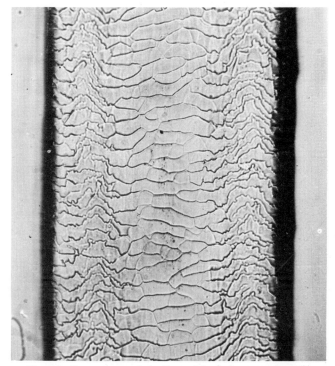

Fig. 20. Common goat hair. Cast of the scale pattern × 250.

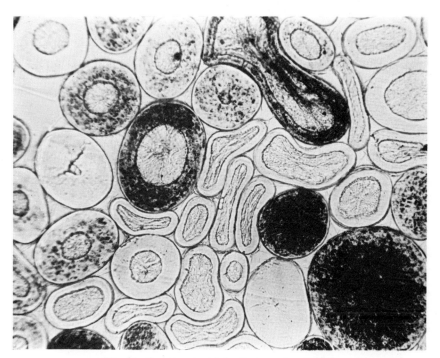

Fig. 21. Common goat hair. Cross-section × 500.

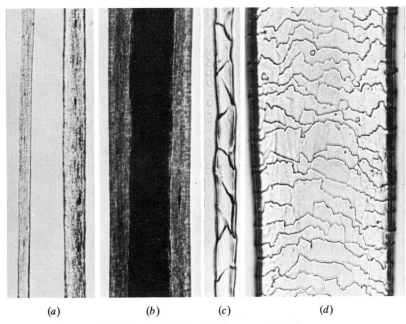

(a)   (b)   (c)   (d)

Fig. 22. Camel hairs. Whole mounts × 180.
(a) Fine fibres. (b) Coarse fibre.
Casts of the scale patterns × 500.
(c) Fine fibre. (d) Coarse fibre.

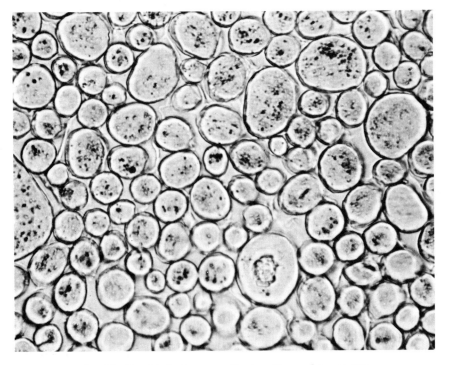

Fig. 23. Camel hairs from superfine top. Cross-section × 500.

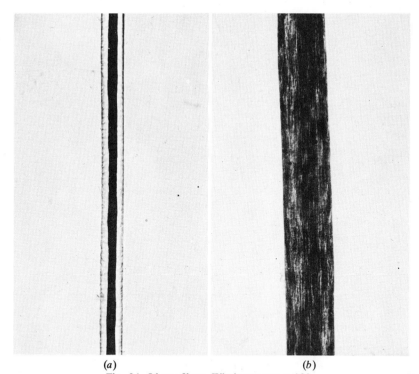

(a)   (b)
Fig. 24. Llama fibres. Whole mounts × 180.
(a) White llama fibre.   (b) Pigmented llama fibre.

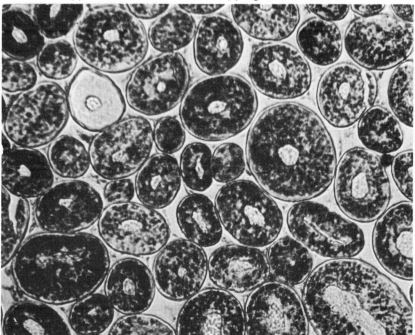

Fig. 25. Llama pigmented fibres. Cross-section × 500. Note the thick cuticle and bi-partite medulla in some fibres.

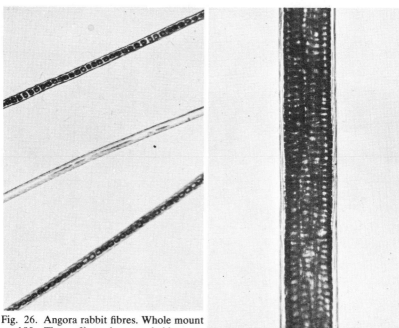

Fig. 26. Angora rabbit fibres. Whole mount × 180. These fibres have a ladder-type medulla that is common to many fur fibres. The fine fibres shown here have a uni-serial medulla, but coarse fibres have a multi-serial medulla similar to the fibres in Fig. 27.

Fig. 27. Hare fibre. Whole mount × 180.

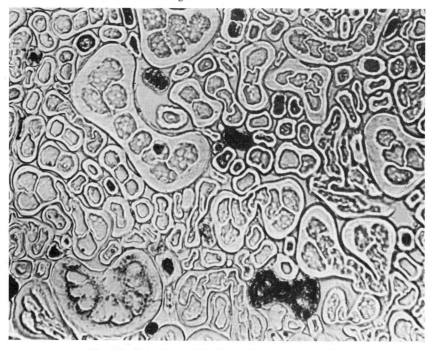

Fig. 28. Light fawn rabbit fibres. Cross-section × 500.

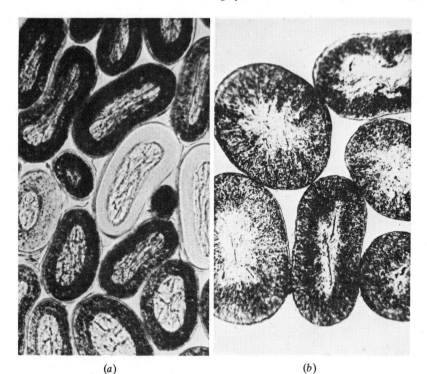

(a)   (b)
Fig. 29. Horse hair. Cross-sections.
(a) Body hair × 250.   (b) Tail hair × 125.
Note the differences in type of medulla between body, mane, and tail hairs.(See also Fig. 30.)

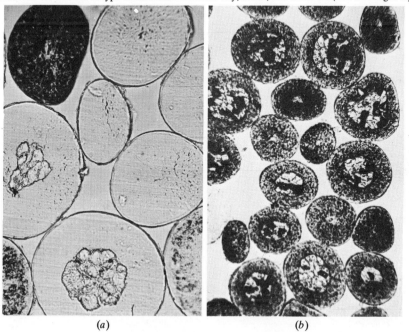

(a)   (b)
Fig. 30. Horse hair. Cross-sections.
(a) White mane hair × 250.   (b) Mane hair × 125.

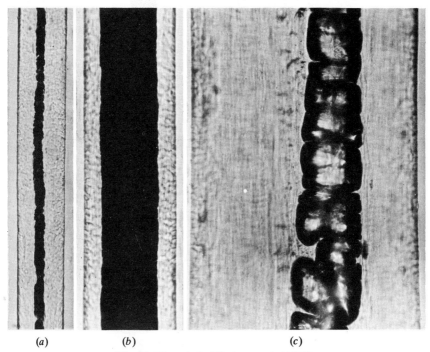

(a)   (b)   (c)

Fig. 31. Horse hair. Whole mounts × 180.
(a) and (b) Mane hair.   (c) Tail hair.

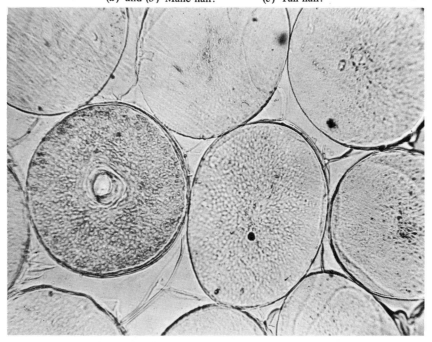

Fig. 32. Cow hairs, adult tail hairs. Cross-section × 250.

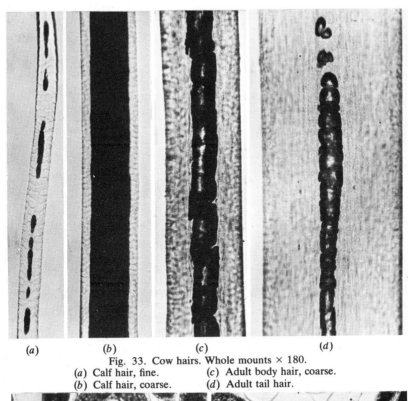

(a)   (b)   (c)   (d)

Fig. 33. Cow hairs. Whole mounts × 180.

(a) Calf hair, fine.   (c) Adult body hair, coarse.
(b) Calf hair, coarse.   (d) Adult tail hair.

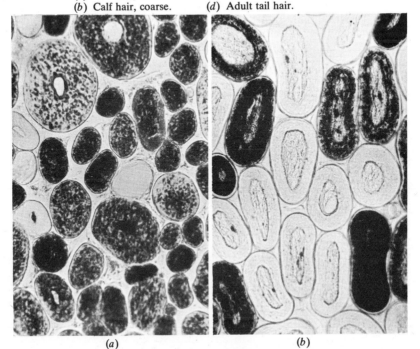

(a)   (b)

Fig. 34. Cow body hairs. Cross-sections × 250.

(a) Calf.   (b) Adult.

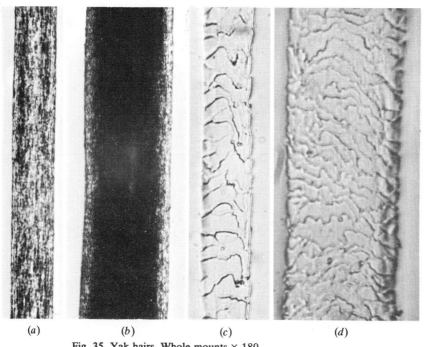

(a)   (b)   (c)   (d)

Fig. 35. Yak hairs. Whole mounts × 180.
(a) Fine fibre. (b) Coarse fibre.
Casts of scale patterns × 500.
(c) Fine fibre, towards the root. (d) Coarse fibre.

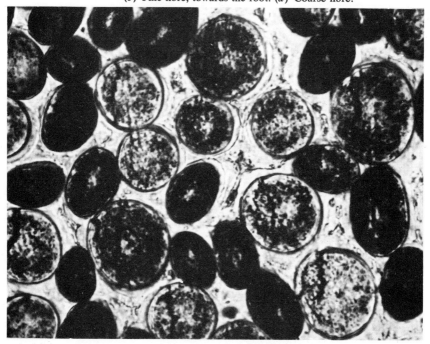

Fig. 36. Yak hairs. Cross-section × 240.

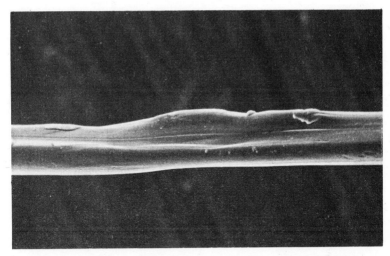

Fig. 37. Silk, *Bombyx mori,* degummed. Whole mount × 1100. SEM.

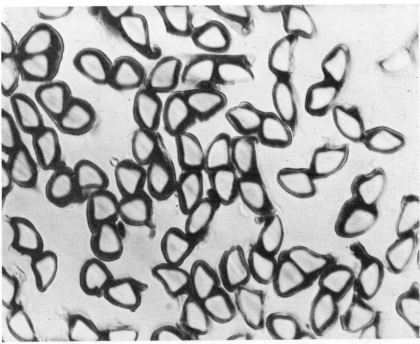

Fig. 38. Silk, *Bombyx mori,* raw. Cross-section × 500. The silk fibres occur in pairs and are held together by the gum which has been stained to differentiate it from the silk.

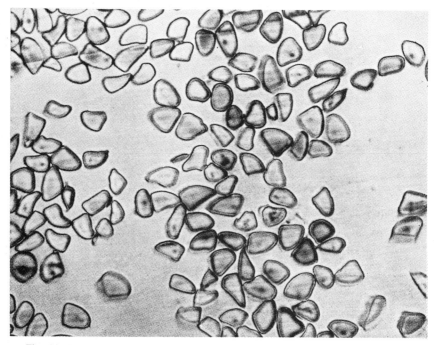

Fig. 39. Silk, *Bombyx mori,* degummed. Cross-section × 500. Compare with Fig. 38.

Fig. 40. Silk, *Bombyx mori,* degummed, lousy. Whole mount × 145.

Fig. 41. Silk, *Bombyx mori*, degummed, lousy. Whole mount × 240. SEM.

Fig. 42. Silk, *Bombyx mori*, degummed, abraded fabric. Whole mount × 145. SEM.

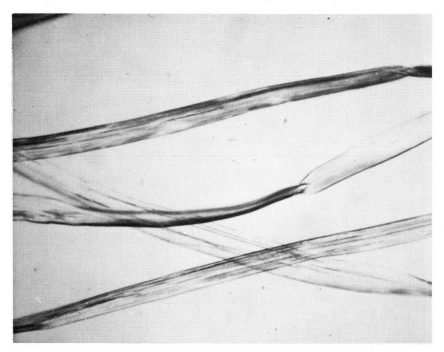

Fig. 43. Silk, wild, Tussah. Whole mount × 200. The striated appearance arises mainly from the internal structure of the flat, ribbon-like fibres.

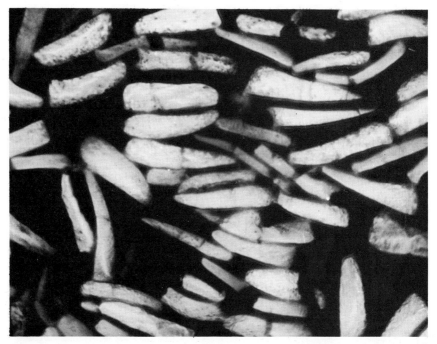

Fig. 44. Silk, wild, Tussah. Ground section × 800. Note the indication of the internal heterogeneities which give the striated appearance to the fibres in Fig. 43.

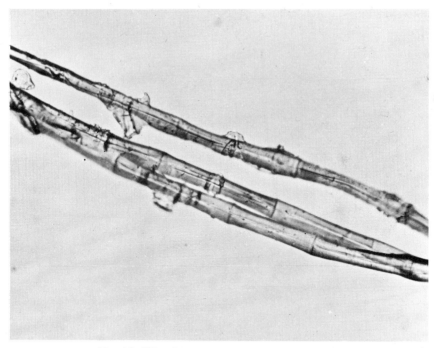

Fig. 45. Silk, wild, Anaphe. Whole mount × 180.

Fig. 46. Silk, wild, Anaphe. Cross-section × 500.

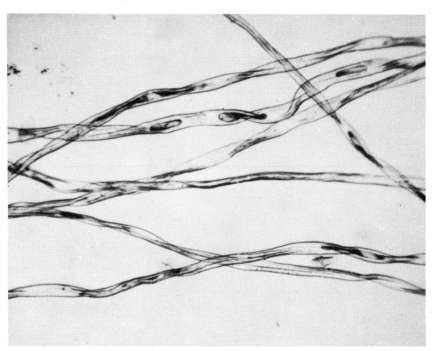

Fig. 47. Cotton, raw, Tanguis. Whole mount × 200. These fibres show more or less pronounced convolutions.

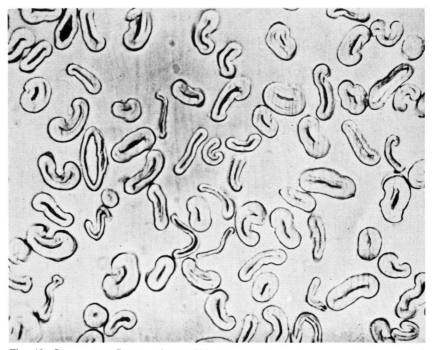

Fig. 48. Cotton, raw. Cross-section × 500. A large variation in shape and size of cross-section is seen. Large lumina are clearly seen.

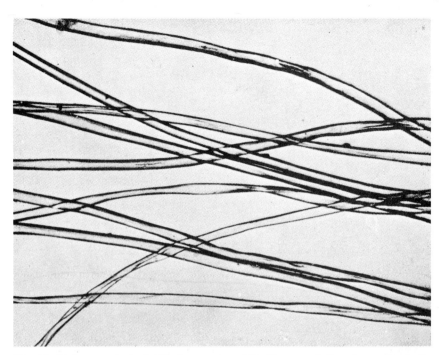

Fig. 49. Cotton, mercerized. Whole mount × 180. The convolutions have almost disappeared. Compare with Fig. 47.

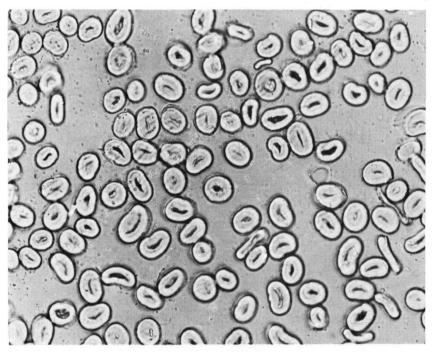

Fig. 50. Cotton, mercerized. Cross-section × 500. Most of the fibres have sections approaching circularity and contain small lumina compared with those of raw cotton seen in Fig. 48.

F

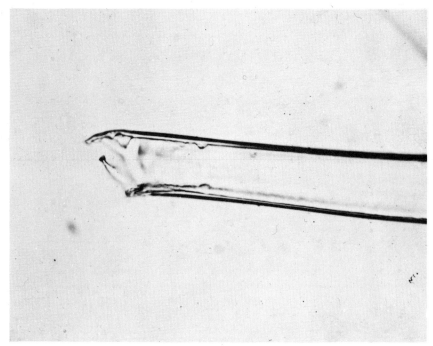

Fig. 51. Akund, base of fibre. Whole mount × 500.

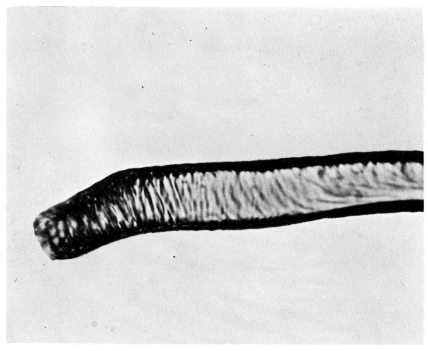

Fig. 52. Kapok, base of fibre. Whole mount × 500.

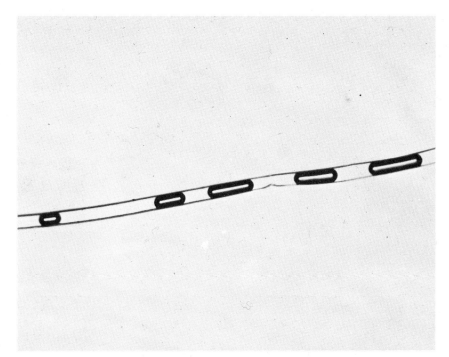

Fig. 53. Kapok. Whole mount × 180. The fibre is very thin-walled as can be seen from the bubbles in the liquid which fills the lumen, and also in the cross-section, Fig. 54.

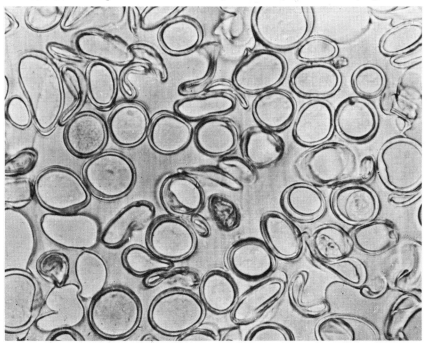

Fig. 54. Kapok. Cross-section × 500.

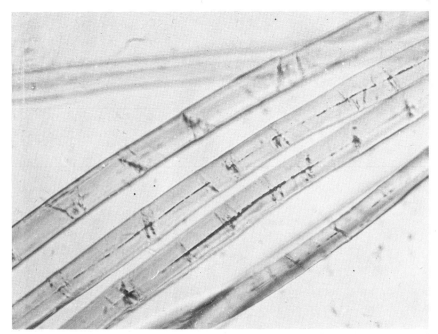

Fig. 55. Flax (ultimates). Whole mount × 500. Note the transverse dislocation marks at frequent intervals along each fibre. These separated ultimate fibres are typical of cottonized flax; alternatively, flax may be processed to retain the ultimate fibres largely in bundles (see Fig. 56).

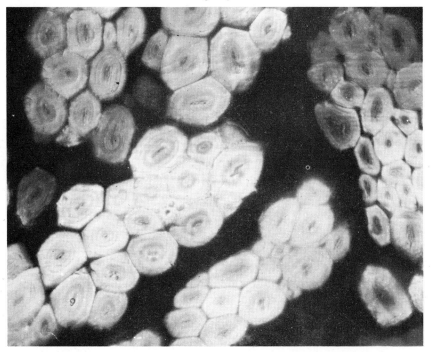

Fig. 56. Flax (bundles). Plate section × 500. The fibre ultimates are grouped together in the bundles in which they grow in the bast. Note the polygonal outlines of the ultimate fibre.

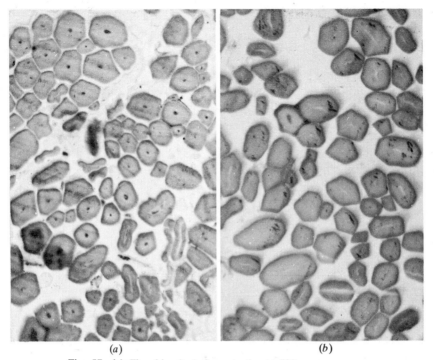

(a)   (b)

Fig. 57. (a) Flax, bleached. Cross-section × 500.
(b) As (a), but mercerized. The lumina are less evident.

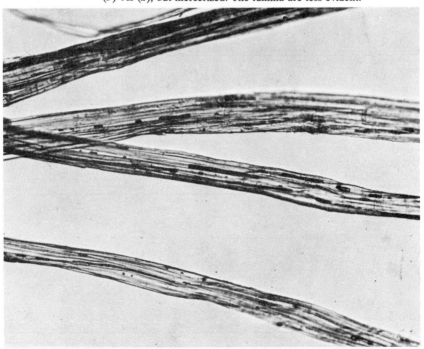

Fig. 58. Jute (bundles). Whole mount × 180. This shows the ultimate fibres present in the bundles in which they grow, and in which they remain in the processed fibre. Compare with Fig. 59.

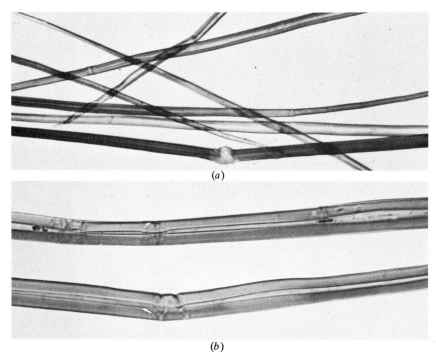

(a)

(b)

Fig. 59. Jute ultimates. Whole mounts.
(a) × 180.   (b) × 500. Note the variation in lumen diameter.

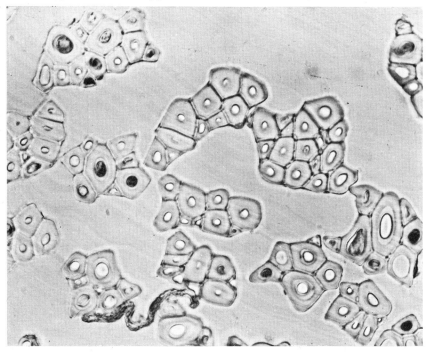

Fig. 60. Jute (bundles). Cross-section × 500. This shows the naturally-occurring bundles of ultimate fibres. Note the polygonal outlines of the ultimates, and note also the circular or elliptical lumina.

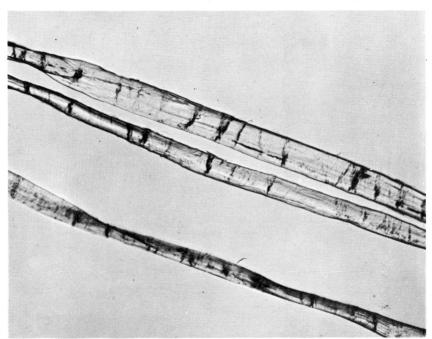

Fig. 61. Hemp (ultimates). Whole mount × 180. Note the transverse dislocation marks at frequent intervals along each fibre. Note also that these are ultimate fibres; hemp is normally processed to retain the ultimate fibres largely in bundles.

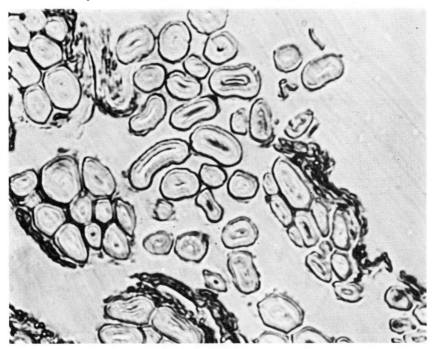

Fig. 62. Hemp (bundles). Cross-section × 340. Many of the fibre ultimates are grouped together in the bundles in which they grow in the bast.

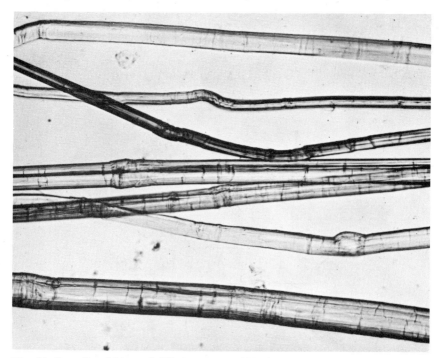

Fig. 63. Sunn fibre (ultimates). Whole mount × 180. Note the cross-markings that appear on the surface of the ultimate fibres.

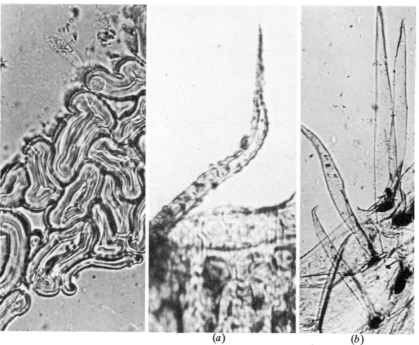

(a)   (b)

Fig. 64. Sunn fibre (bundle). Cross-section × 500. Note the irregular outline of the ulti-mates.

Fig. 65. Fibre debris (see pp. 18 and 231). Whole mounts of epidermis.
    (a) Hemp × 400. Note dark cell contents and single tapered hair with small warts (top left).
    (b) Sunn fibre × 200. Note numerous smooth hairs.

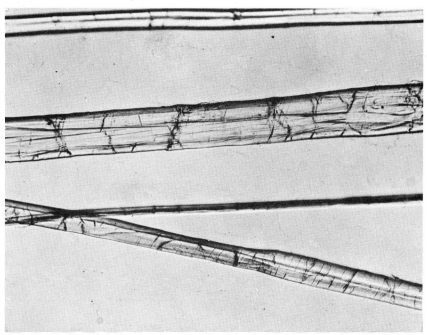

Fig. 66. Ramie (ultimates). Whole mount × 180. Note the transverse dislocation marks present at frequent intervals along each fibre. Note also that these are ultimate fibres separated from the bundles. This separation also takes place during the commercial degumming prior to spinning (see p. 18).

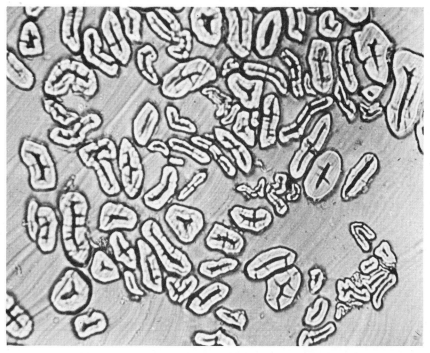

Fig. 67. Ramie, cross-section × 340. Note the tendency for radial cracks to develop in these fibres.

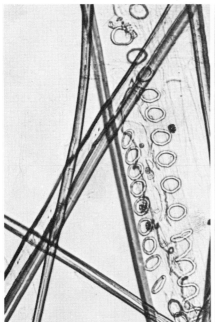 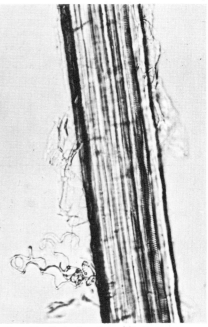

Fig. 68. Sisal (ultimates). Whole mount × 180. Here the ultimates have been separated from the bundles which are seen in Fig. 69. Note the spiral and annular vessels present.

Fig. 69. Sisal (bundle). Whole mount × 180. This is part of a bundle made up of ultimate fibres. Note the threads from the spiral vessels.

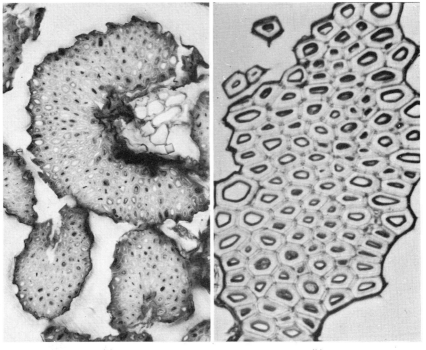

(a)   (b)

Fig. 70. (a) Sisal (bundles). Ground section (see Section 4.1.5.4) × 180. Note the crescent-shaped outline.
    (b) Sisal (bundles). Cross-section × 500. Note the polygonal outline of the ultimate fibres.

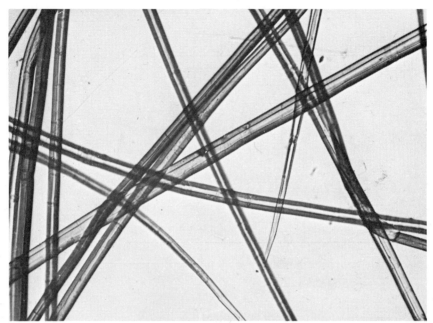

Fig. 71. Abaca or Manila (ultimates). Whole mount × 180. Here the ultimates have been separated from the fibre bundles.

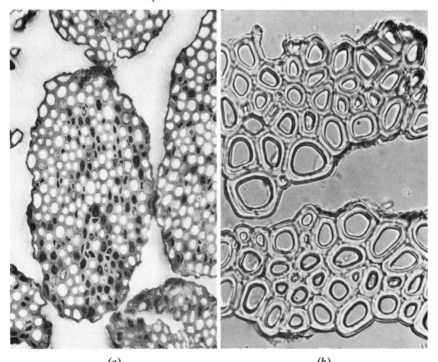

(a)   (b)

Fig. 72. (a) Abaca (bundles). Ground section (see Section 4.1.5.4) × 180. Note the oval outlines.

(b) Abaca (bundles). Cross-section × 500. Note the polygonal ultimates.

Fig. 73. *Phormium tenax* (ultimates). Whole mount × 180. Here the ultimates have been separated from the fibre bundles. Note the spiral vessels.

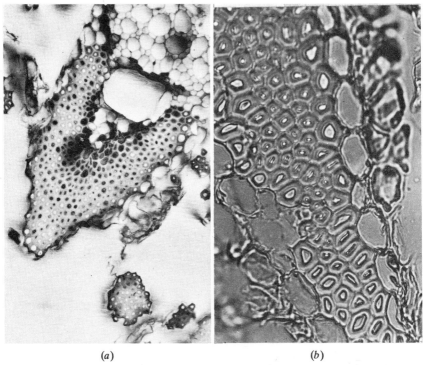

(a)   (b)

Fig. 74. (a) *Phormium tenax* (bundles). Ground section (see Section 4.1.5.4) × 180. Note the crescent-shaped outlines.
(b) *Phormium tenax* (bundles). Cross-section × 500. Note the polygonal ultimates with small lumina.

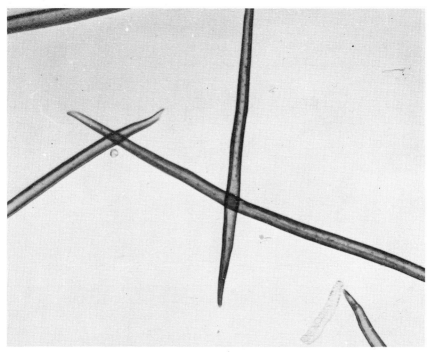

Fig. 75. Coir (ultimates). Whole mount × 180. Here the ultimates have been separated from the fibre bundles.

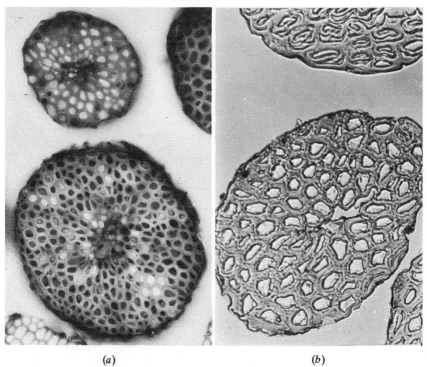

(a)   (b)

Fig. 76. (a) Coir (bundles). Ground section (see Section 4.1.5.4) × 180. Note the almost elliptical outline of the fibre bundle.
 (b) Coir (bundles). Cross-section × 500. These fibre bundles are considerably finer than those shown in the ground section.

(a)

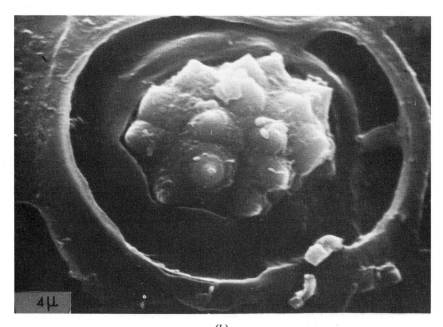

(b)
Fig. 77. Coir fibre. Whole mounts. SEM.
(a) Surface view showing siliceous stegmata. (b) A single stegmatum.

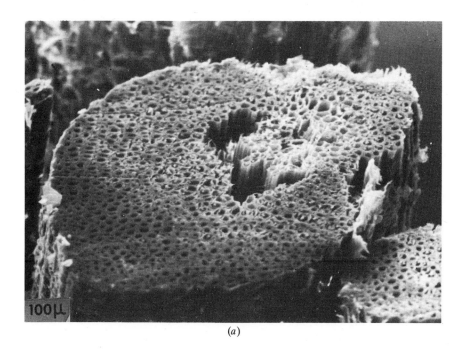

(*a*)

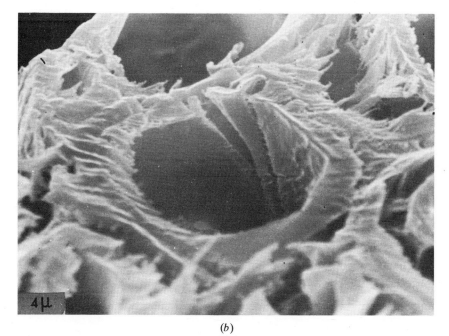

(*b*)

Fig. 78. (*a*) Coir (bundle). Cross-section showing ultimate fibre cells forming fibre bundle. SEM.
(*b*) As (*a*) at higher magnification. SEM.

(a)   (b)

Fig. 79. (a) Jute × 900. Ash showing solitary crystals mounted in water.
(b) *Urena lobata* × 900. Ash showing cluster crystals mounted in water.

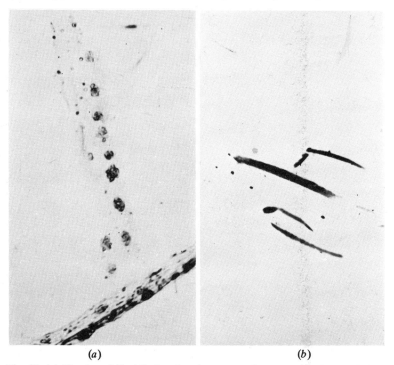

(a)   (b)

Fig. 80. (a) Hemp × 340. Ash showing cluster crystals mounted in Canada balsam.
(b) Sisal × 340. Ash showing rod-like crystals mounted in aniline.

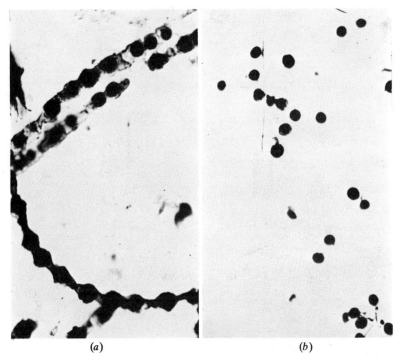

(a)   (b)

Fig. 81. (a) Abaca or Manila × 340. Ash showing stegmata mounted in clove oil and phenol.
(b) Coir × 340. Ash showing stegmata mounted in clove oil and phenol.

Fig. 82. Asbestos, amosite. Whole mount. TEM.

Fig. 83. Asbestos, crocidolite. Whole mount. TEM.

Fig. 84. Asbestos, chrysotile. Whole mount. TEM. Note the fibrillar structure.

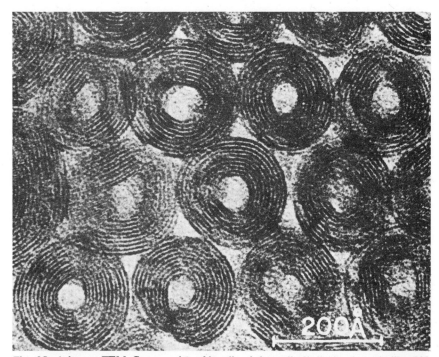

Fig. 85. Asbestos. TEM. Cross-section of bundle of chrysotile fibrils showing spiralling 002 planes.

Fig. 86. Asbestos, amosite. Electron diffraction pattern.

Fig. 87. Asbestos, crocidolite. Electron diffraction pattern.

Fig. 88. Asbestos, chrysotile. Electron diffraction pattern.

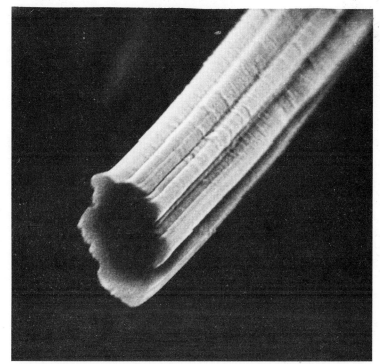

Fig. 89. Regenerated viscose fibre (Fibro). Whole mount × 2200. SEM. Numerous striations are a feature of regular (normal-tenacity) regenerated viscose.

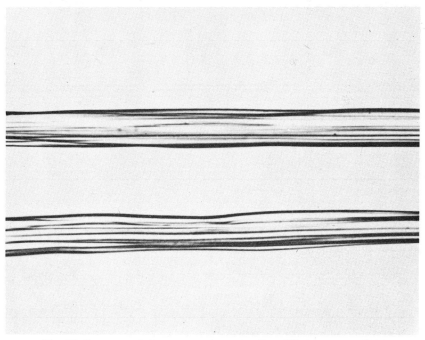

Fig. 90. Regenerated viscose, normal-tenacity fibre. Whole mount × 500.
Upper fibre—Continuous-filament yarn.
Lower fibre—Fibro staple fibre.

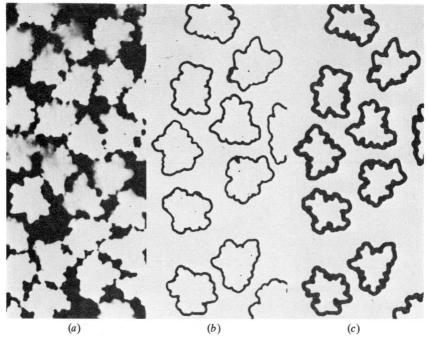

(a)   (b)   (c)

Fig. 91. Regenerated viscose, normal-tenacity (continuous-filament yarn). Cross-sections
× 500. The serrated outline of the filament is characteristic.
(a) Plate section.   (b) Thin section.   (c) Skin-stained section.

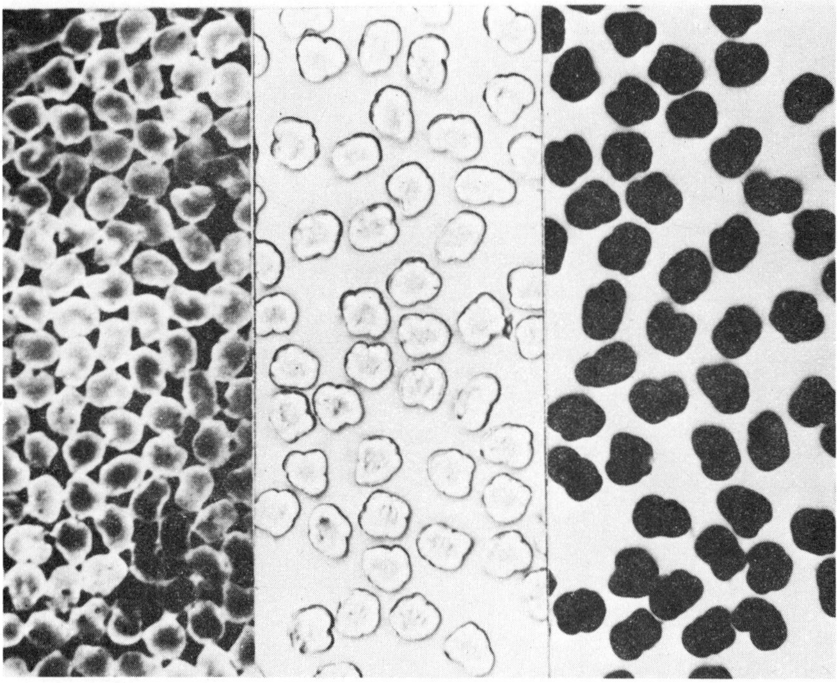

(a)   (b)   (c)

Fig. 92. Regenerated viscose, high-tenacity continuous-filament tyre yarn (Tenasco Super
2A). Cross-sections × 500. Cross-sectional outline circular to bean-shaped.
(a) Plate section.   (b) Thin section.   (c) Skin-stained section.

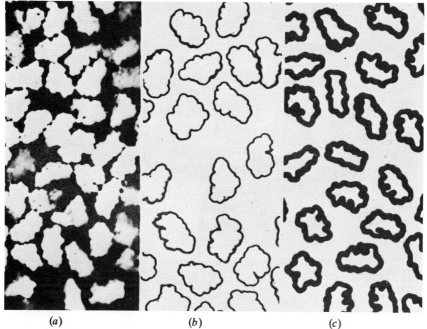

(a)   (b)   (c)

Fig. 93. Regenerated viscose, staple fibre (Fibro). Cross-sections × 500. The serrated outline of the fibres is characteristic; the fibre outline may be slightly elongated.
(a) Plate section.   (b) Thin section.   (c) Skin-stained section.

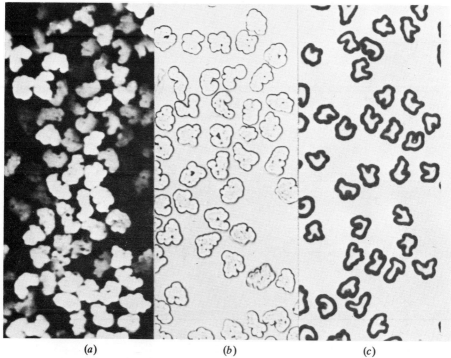

(a)   (b)   (c)

Fig. 94. Regenerated viscose, improved-tenacity staple fibre. Cross-sections × 500. Note
fibre outline is smoothly indented rather than serrated, with a tendency towards a 'C' shape.
(a) Plate section.   (b) Thin section.   (c) Skin-stained section.

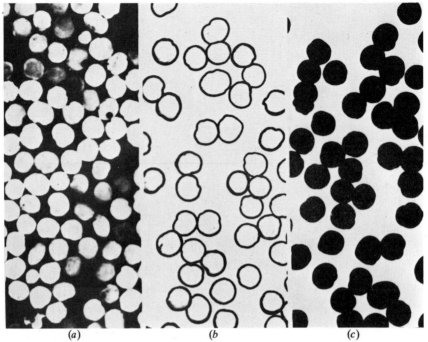

Fig. 95. Regenerated viscose, high-tenacity staple fibre (Durafil—now obsolete). Cross-
sections × 500. Cross-sectional outline mainly circular.
(*a*) Plate section.   (*b*) Thin section.   (*c*) Skin-stained section.

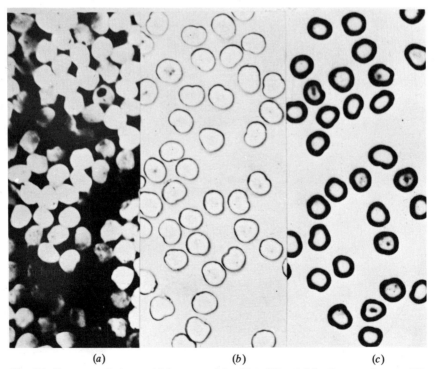

(*a*)   (*b*)   (*c*)

Fig. 96. Regenerated viscose, high-wet-modulus fibre (Vincel 64). Cross-sections × 500.
Cross-sectional outline circular or slightly off-round to bean-shaped.
(*a*) Plate section.   (*b*) Thin section.   (*c*) Skin-stained section.

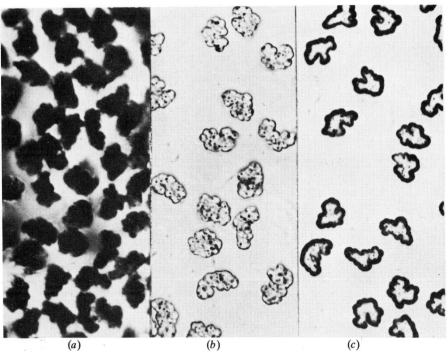

(a)   (b)   (c)

Fig. 97. Regenerated viscose, crimped fine fibre (Sarille). Cross-sections × 500. Presence of
uneven skin thickness shown in stained section.
(a) Plate section.   (b) Thin section.   (c) Skin-stained section.

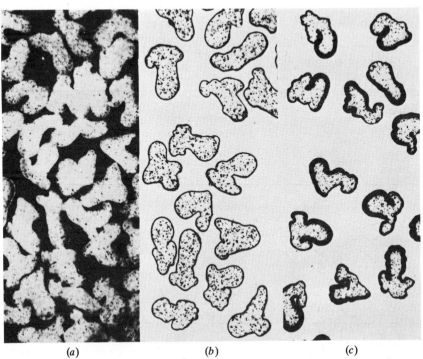

(a)   (b)   (c)

Fig. 98. Regenerated viscose, crimped and modified coarse fibre used, e.g., in carpets
(Evlan). Cross-sections × 250. Uneven skin thickness; this staple has a smoother fibre outline
than an unmodified staple.
(a) Plate section.   (b) Thin section.   (c) Skin-stained section.

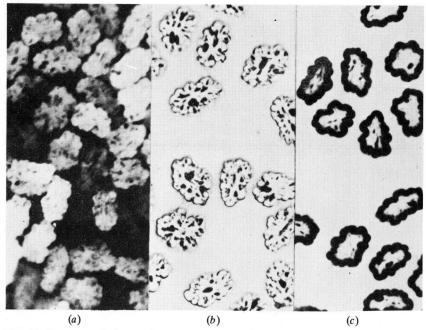

(a)   (b)   (c)

Fig. 99. Regenerated viscose, flame-retardant fibre (Darelle). Cross-sections × 500. This sample is regular viscose type; F.R. fibres can also show crimped characteristics as in Figs 97 and 98.

(a) Plate section.   (b) Thin section.   (c) Skin-stained section.

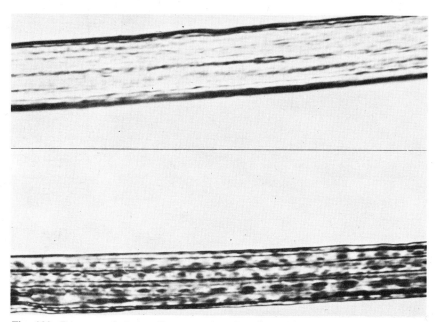

Fig. 100. Regenerated viscose, flame-retardant fibre (Darelle). Whole mount × 500. Note globules of F.R. additive in fibre. Lower specimen shows globules stained with Celliton Blue FFR.

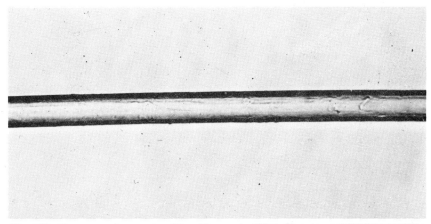

Fig. 101. Cuprammonium rayon fibre (Cupresa). Whole mount × 750. The rod-like appearance is in contrast to that of regenerated viscose (see Fig. 90).

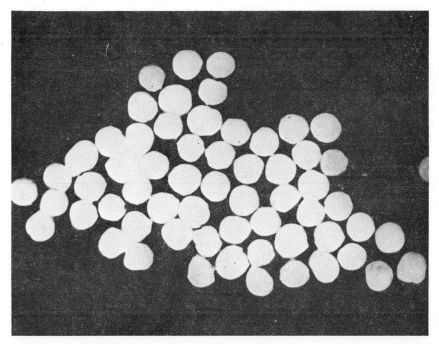

Fig. 102. Cuprammonium rayon fibre (Cupresa). Cross-section × 750. The cross-section shows roundish shapes with occasional flattening and fusion.

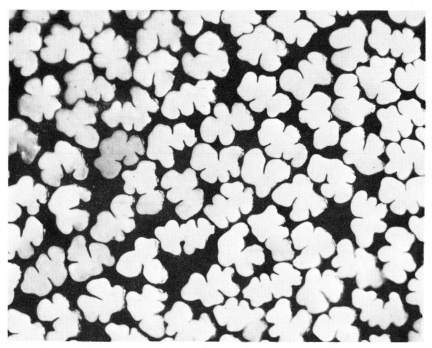

Fig. 103. Cellulose diacetate fibre (Dicel). Plate section × 500. The lobes on each section are fewer than for regenerated viscose, as seen in Fig. 91.

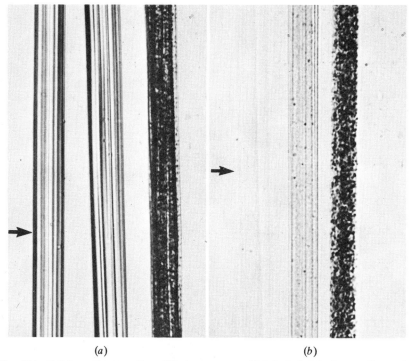

(a)   (b)

Fig. 104. Cellulose diacetate fibre. Whole mounts × 300. Left, bright (arrowed); centre, soap delustred; right, matt.
(a) Mounted in water.   (b) Mounted in white mineral oil.

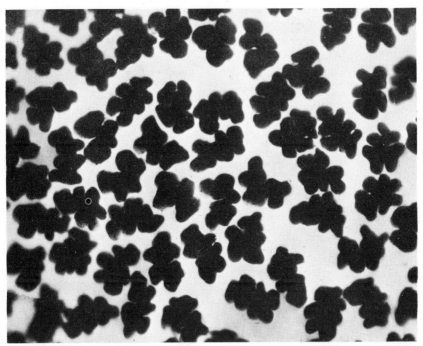

Fig. 105. Cellulose diacetate, matt fibre. Plate section mounted in 1–bromonaphthalene × 500. The presence of the high-refractive-index medium causes light to be transmitted preferentially through the spaces between fibres.

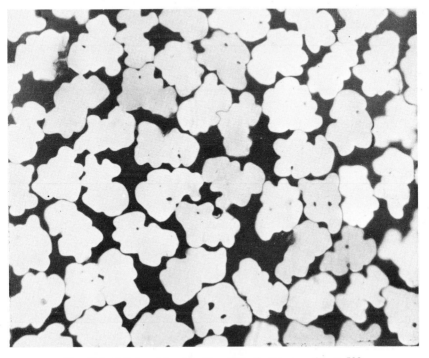

Fig. 106. Cellulose triacetate fibre (Tricel). Cross-section × 500.

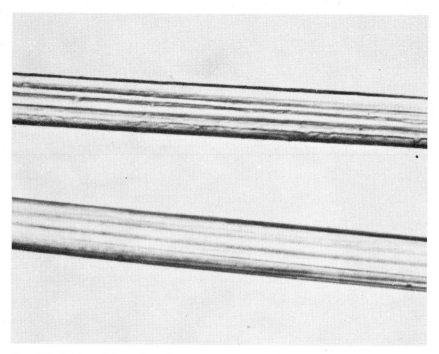

Fig. 107. Calcium alginate fibre (Courtaulds' Alginate). Whole mount × 500. Ribbon-like behaviour and fine striations are characteristic features.

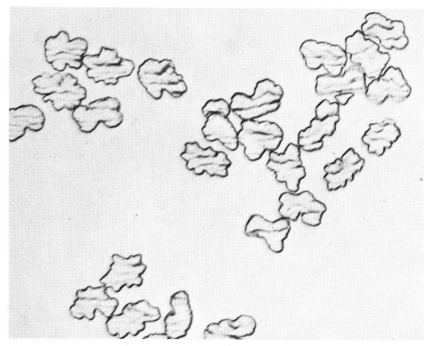

Fig. 108. Calcium alginate fibre (Courtaulds' Alginate). Cross-section cut from fibres embedded in gelatin × 500. The section outlines are elongated and notched, with numerous irregular serrations.

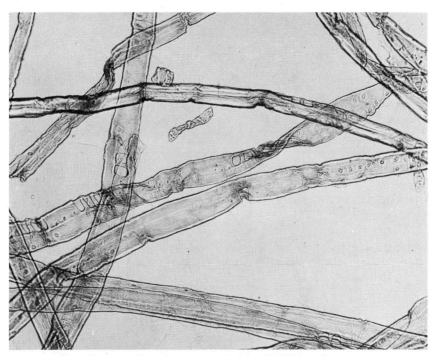

Fig. 109. Paper fibres (Pine 'Kraft' Sulphate). Whole mount × 200.

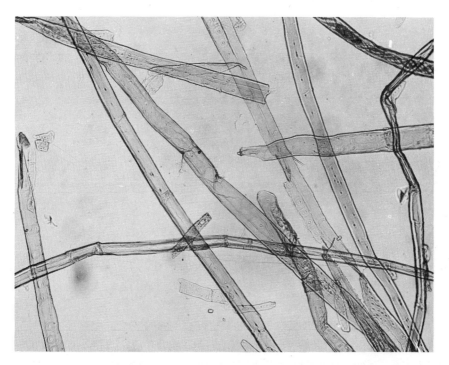

Fig. 110. Paper fibres (Spruce Sulphite). Whole mount × 200.

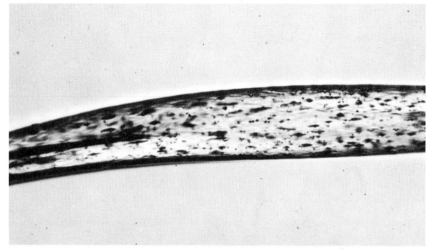

Fig. 111. Acrylic fibre (Orlon 42). Whole mount × 750. The ribbon-like twist that is indicated is characteristic of fibres with a dogbone-like or elongated cross-sectional shape (see Fig. 112).

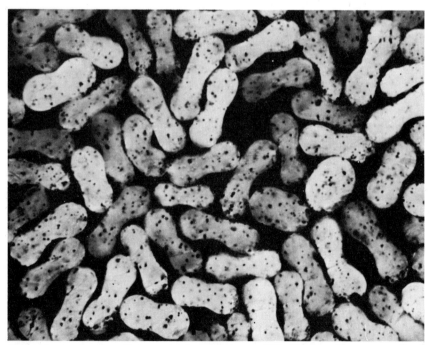

Fig. 112. Acrylic fibre (Orlon 42). Cross-section × 750. Dogbone-like cross-sectional shape.

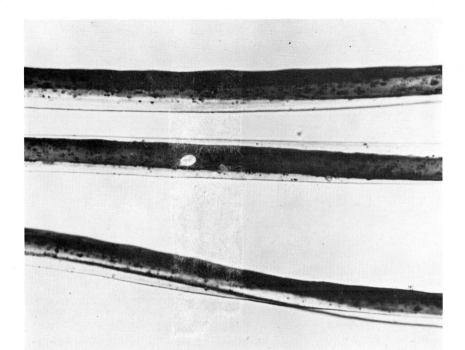

Fig. 113. Acrylic fibre with bicomponent structure (Orlon 21 or Sayelle). Whole mount × 500. Stained in Shirlastain E, which develops a red coloration on one side of the fibre, the other side remaining unstained.

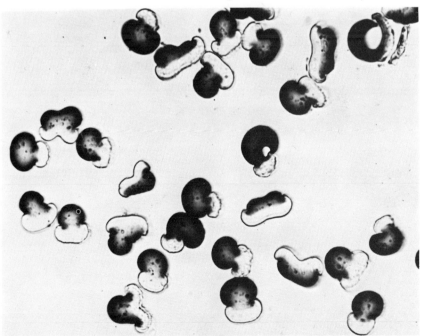

Fig. 114. Acrylic fibre with bicomponent structure (Orlon 21 or Sayelle). Thin cross-section cut from stained yarn embedded in n-butyl methacrylate × 500. Lobed cross-sectional shape.

H

Fig. 115. Acrylic fibre (Courtelle). Whole mount × 1100. SEM. The very fine streaks on the fibre surface are characteristic of Courtelle. (See also Fig. 116.)

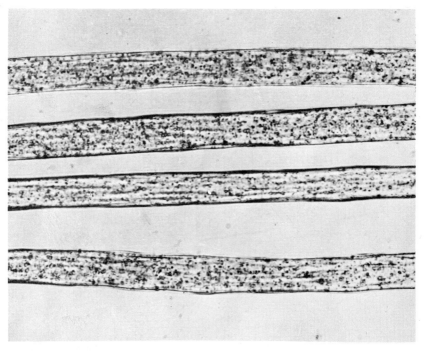

Fig. 116. Acrylic fibre (Courtelle). Whole mount × 500.

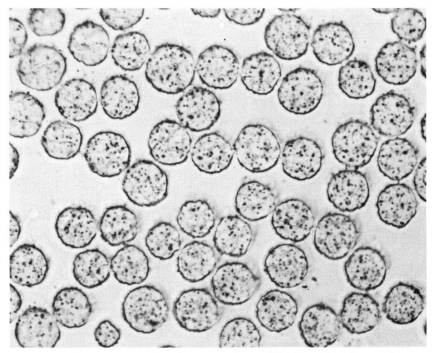

Fig. 117. Acrylic fibre (Courtelle). Cross-section × 625. Approximately round cross-sectional shape, with extremely fine indentations around the periphery only visible at higher magnifications.

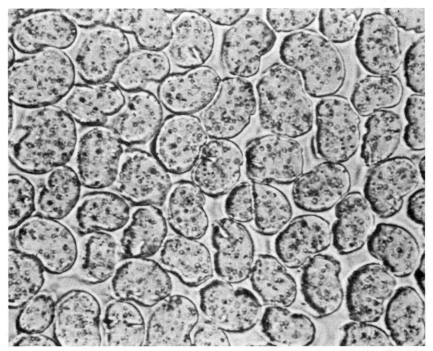

Fig. 118. Acrylic fibre (Acrilan 16). Cross-section × 600. Cross-sectional shapes off-round. Compare this view of 3·3 dtex fibres with that of 17 dtex fibres in Fig. 119.

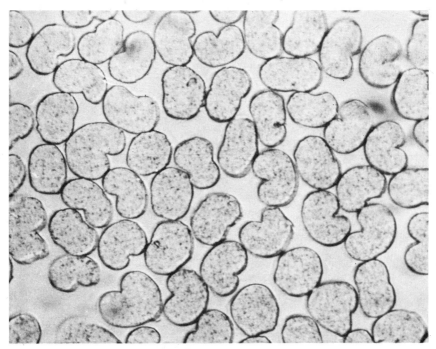

Fig. 119. Acrylic fibre (Acrilan, 17 dtex). Cross-section × 240.

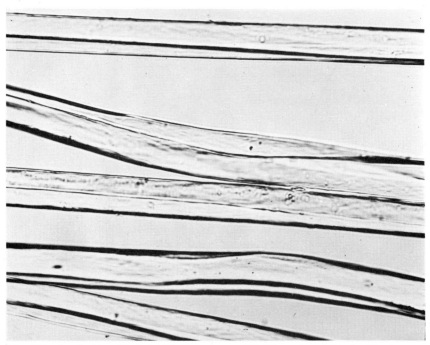

Fig. 120. Acrylic fibre (Crylor). Whole mount × 500. The twists visible in this view are typical of a fibre with an unbalanced cross-sectional outline (see Fig. 121).

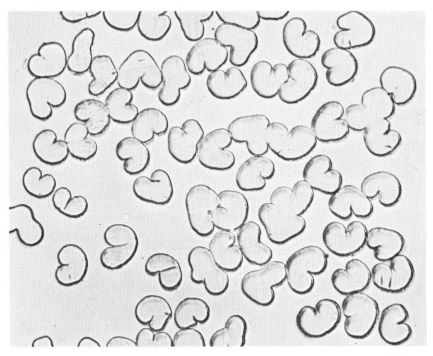

Fig. 121. Acrylic fibre (Crylor). Methacrylate section × 500. Lobed cross-sectional shape.

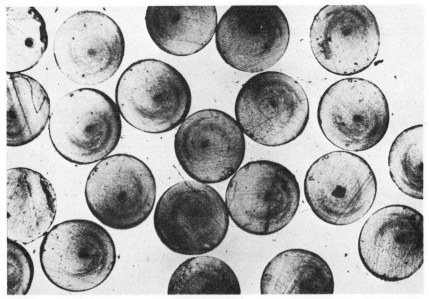

Fig. 122. Poly(vinylidene chloride) fibre (Saran). Cross-section × 75. The markings within the fibres are due to section preparation and photography. The fibres are featureless rods in longitudinal view.

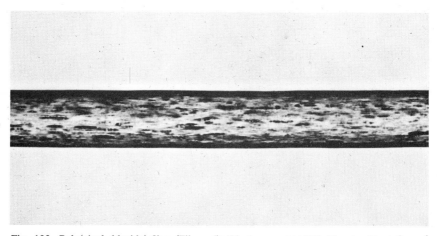

Fig. 123. Poly(vinyl chloride) fibre (Fibravyl). Whole mount × 750. The streaky surface of the fibre is characteristic.

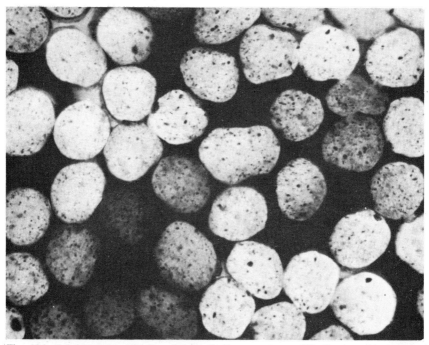

Fig. 124. Poly(vinyl chloride) fibre (Fibravyl). Cross-section × 750. Cross-sectional shapes are irregularly rounded.

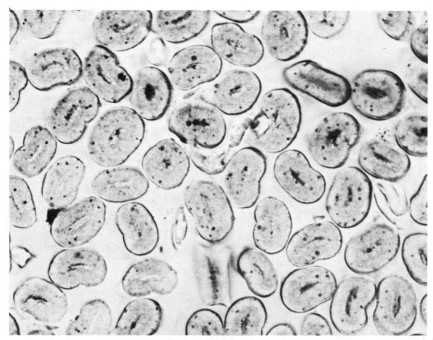

Fig. 125. Syndiotactic poly(vinyl chloride) fibre (Leavil). Cross-section × 800. Note lumen similar to cotton fibres (cf. Fig. 48) and delustrant particles.

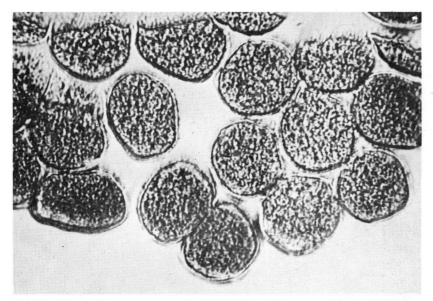

Fig. 126. Polytetrafluoroethylene fibre (Teflon). Cross-section × 700. The fibres are very dull, with a granular appearance.

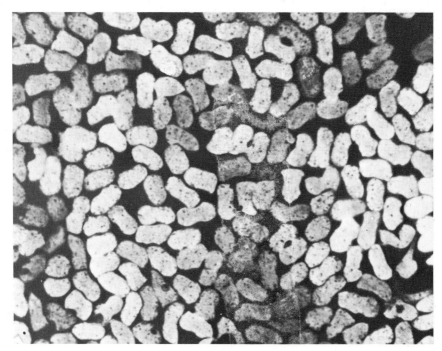

Fig. 127. Modacrylic fibre (Teklan). Cross-section × 500. The irregular shapes in this cross-section contrast with the characteristic, folded outlines of Dynel (see Fig. 129).

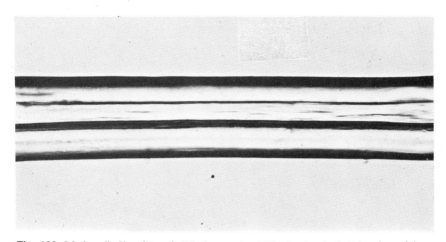

Fig. 128. Modacrylic fibre (Dynel). Whole mount × 750. The deeply-fluted surface of these fibres is characteristic.

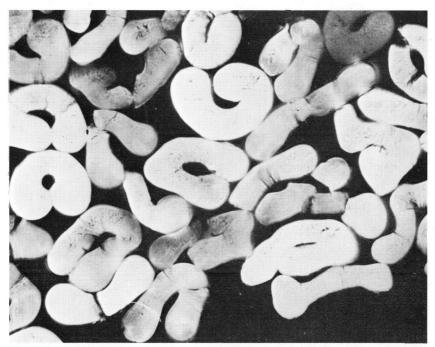

Fig. 129. Modacrylic fibre (Dynel). Cross-section × 750. Folded shapes with deep re-entrants. Some fibres have cracked during the preparation of the section.

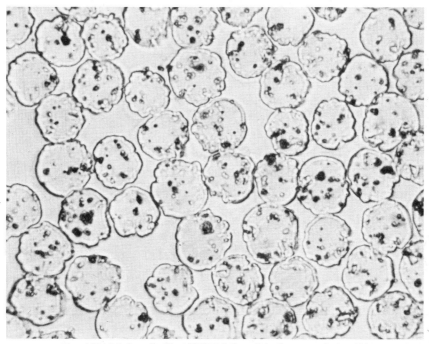

Fig. 130. Modacrylic fibre (SEF Modacrylic). Cross-section × 600.

Fig. 131. Nylon 6.6. Whole mount × 500. This shows the appearance of fibres with three levels of content of titanium dioxide delustrant; left, extra-dull fibre, 2% TiO_2; centre, dull fibre, 0·5% TiO_2; right, bright fibre, 0·03% TiO_2.

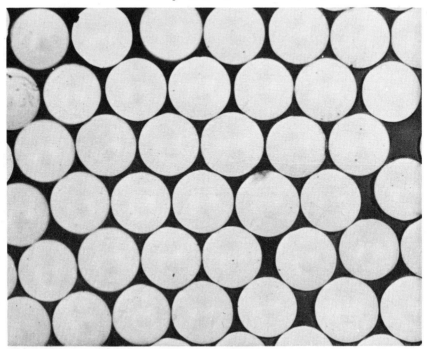

Fig. 132. Nylon 6.6. Cross-section × 400. Other nylon fibres, e.g., nylons 6 and 11, are similar.

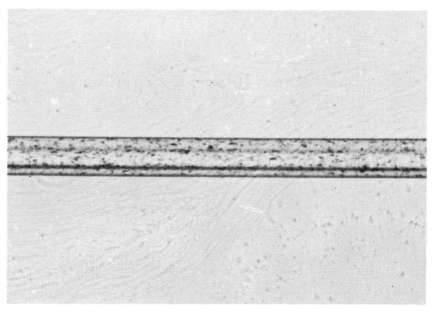

Fig. 133. Nylon 6.6, trilobal. Whole mount × 500.

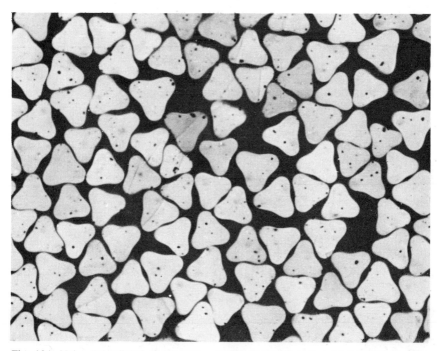

Fig. 134. Nylon 6.6, trilobal. Cross-section × 300. Bright fibre containing 0·05% titanium dioxide.

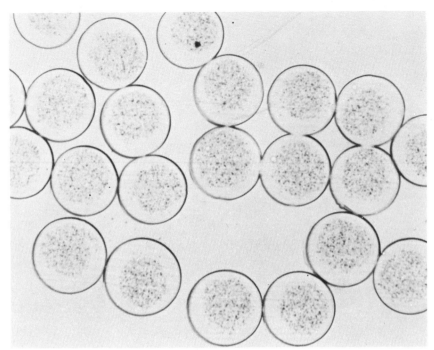

Fig. 135. Bicomponent fibres, sheath–core type. Cross-section × 160. The two nylon components are not normally easy to differentiate under the microscope; in the examples shown, the components differ in level of pigmentation.

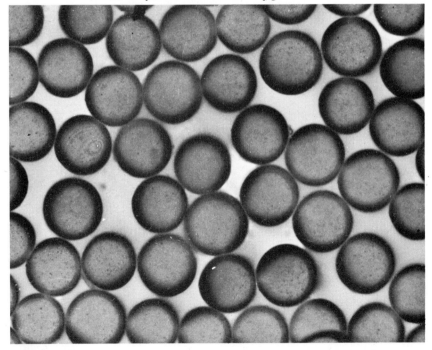

Fig. 136. Bicomponent sheath–core fibres (nylon 6–nylon 6.6). Cross-section × 500. Stained with iodine solution to show distribution of nylon 6 sheath (see Section 5.20).

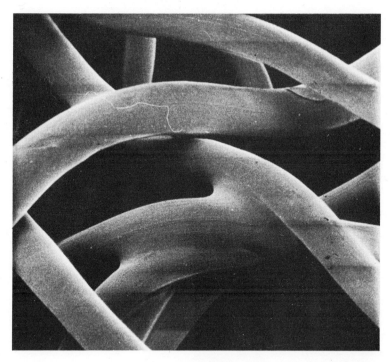

Fig. 137. Bicomponent sheath–core type fibres. Whole mount × 300. SEM. Nylon fibres are fused ('melded') by melting of the sheath component.

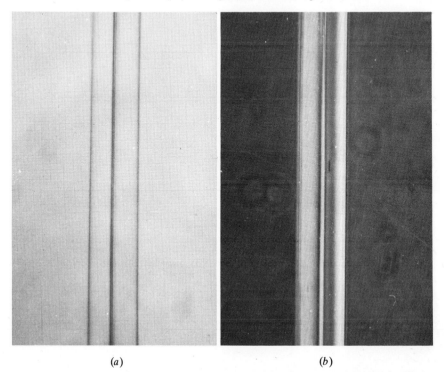

(a)   (b)

Fig. 138. Bicomponent fibres, side-by-side type, nylon–polyurethane (Monvelle). Whole mount × 200.
(a) Seen by ordinary light microscope.   (b) Seen by phase contrast.

Fig. 139. Bicomponent fibres, side-by-side type, nylon–polyurethane (Monvelle). Ground section × 400.

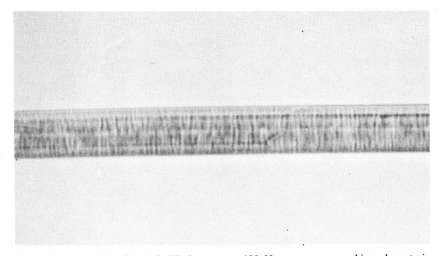

Fig. 140. Aramid fibre (Kevlar). Whole mount × 400. Note transverse markings characteristic of Kevlar.

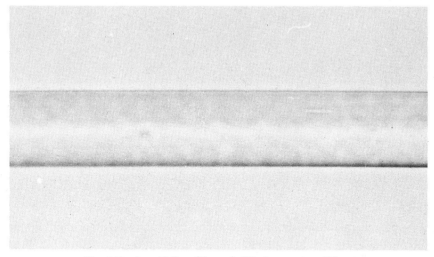

Fig. 141. Aramid fibre (Nomex). Whole mount × 400.

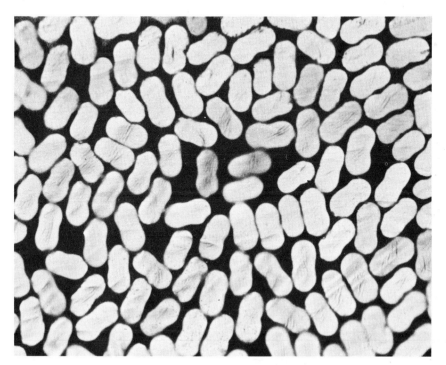

Fig. 142. Aramid fibre (Nomex). Cross-section × 400.

Fig. 143. Aramid fibre (Conex). Whole mount × 400.

Fig. 144. Aramid fibre (Conex). Cross-section × 400.

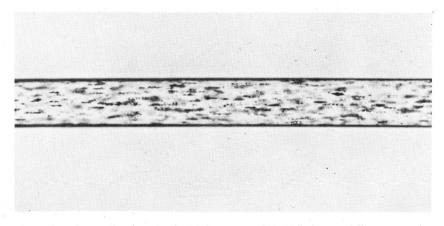

Fig. 145. Polyester fibre (Terylene). Whole mount × 500. Dull pigmented filament containing 0·5% titanium dioxide.

Fig. 146. Polyester fibre (Terylene). Cross-section with black oil paint (see Section 4.1.5.1) × 500. Dull pigmented filament containing 0·5% titanium dioxide.

Fig. 147. Bicomponent polyester fibre, side-by-side type. Cross-section × 160.

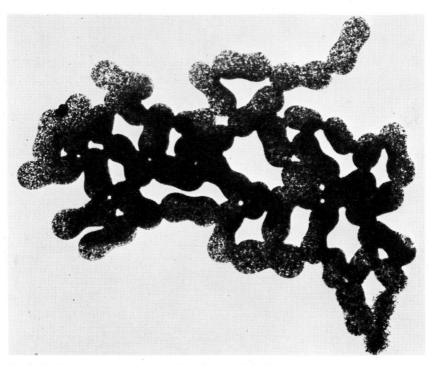

Fig. 148. Polyurethane elastomeric fibre (Lycra 124). Cross-section × 250. Multifilament yarn with fusion of filaments.

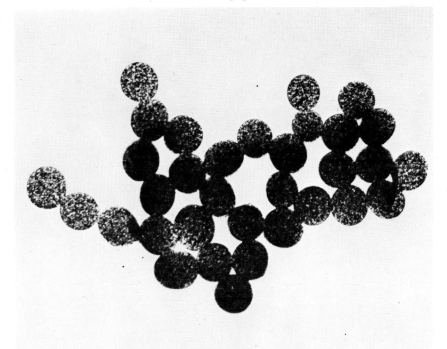

Fig. 149. Polyurethane elastomeric fibre (Lycra 126). Cross-section × 250. Multifilament yarn with fused circular filaments.

Fig. 150. Polyurethane elastomeric fibre (Spanzelle). Cross-section × 350. Multifilament yarn with fusion of filaments.

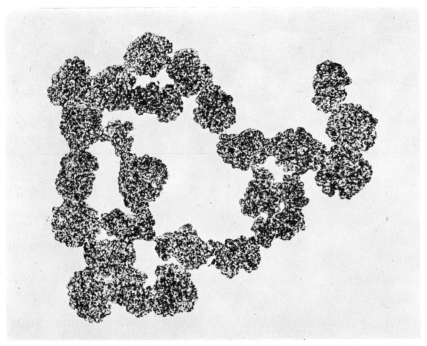

Fig. 151. Polyurethane elastomeric fibre (Sarlane). Cross-section × 250. Multifilament yarn
with fusion of filaments.

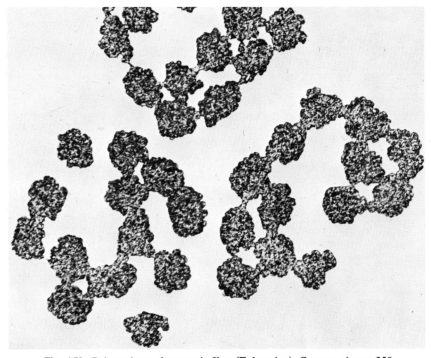

Fig. 152. Polyurethane elastomeric fibre (Enkaswing). Cross-section × 250.

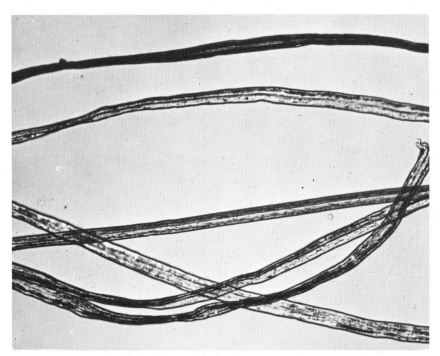

Fig. 153. Poly(vinyl alcohol) fibre (Kuralon). Whole mount × 180. These fibres are more irregular than most man-made fibres.

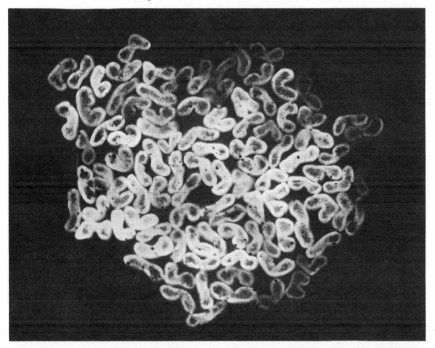

Fig. 154. Poly(vinyl alcohol) fibre (Kuralon). Plate section mounted in n-decane × 500. The fibres show a transparent outer region and a pitted core.

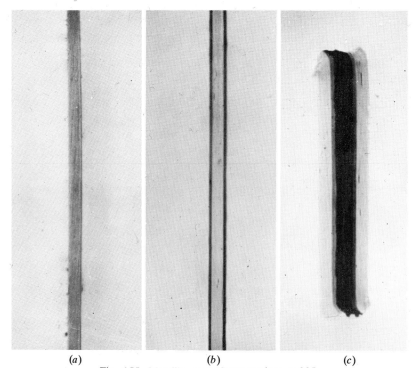

(a)   (b)   (c)

Fig. 155. Metallic yarns. Cross-sections × 285.
(a) Lurex clear N 50.   (b) Lurex white-gold TE 50.   (c) Lurex white-gold MF 150.

Fig. 156. Metallic yarn (Lurex white-gold MF 150). Cross-section × 1150. TEM.

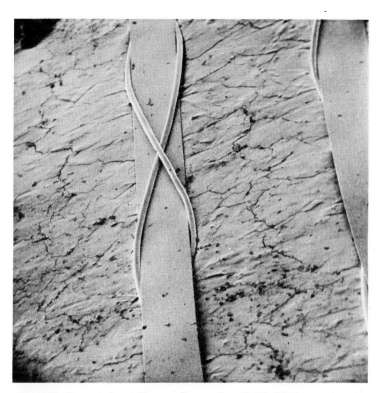

Fig. 157. Supported metallic yarn (Lurex silver C 50). Whole mount × 44.

Fig. 158. Cross-section of fibres mounted in a cross-section stub for examination in the SEM.

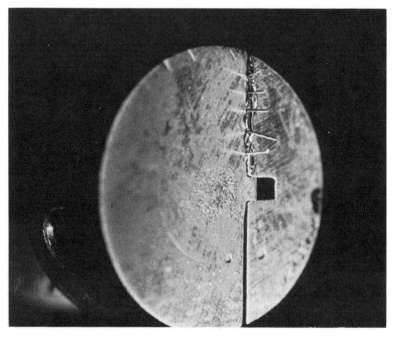

Fig. 159. Cross-section stub being used to examine broken fibre ends in the SEM.

Fig. 160. SEM stub showing single fibres and yarn mounted for examination.

Fig. 161. Fabric sample mounted on an SEM stub for examination.

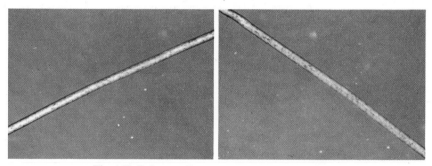

Fig. 162. Polyester fibres seen under polarized light.
(a)n$_{\parallel}$.· (b)n$_{\perp}$.

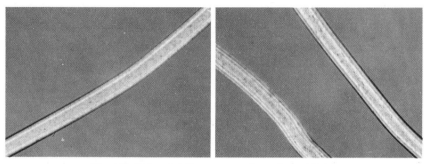

Fig. 163. Nylon 6.6 fibres seen under polarized light.
(a)n$_{\parallel}$. (b)n$_{\perp}$.

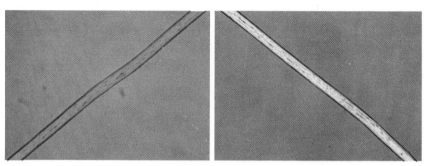

Fig. 164. Poly(vinyl chloride) fibres seen under polarized light.
(a)n$_{\parallel}$. (b)n$_{\perp}$.

Fig. 165. Acrylic fibres seen under polarized light.
(a)n$_{\parallel}$. (b)n$_{\perp}$.

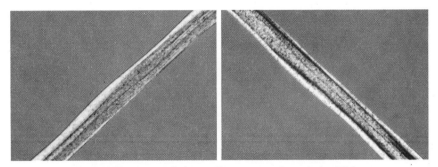

Fig. 166. Regenerated viscose fibres seen under polarized light.
(a)n$_{\parallel}$. (b)n$_{\perp}$.

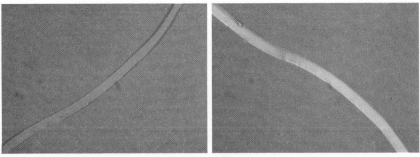

Fig. 167. Cellulose diacetate fibres (Dicel) seen under polarized light.
(a)n$_{\parallel}$. (b)n$_{\perp}$.

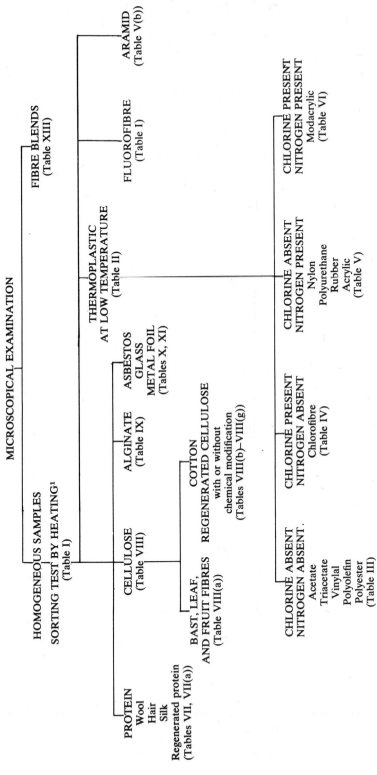

4.1 Microscopical Examination of Fibre Specimens

The task of fibre identification is greatly assisted by a preliminary examination of the sample with a microscope. All fibres have their distinguishing features which, even if not sufficiently characteristic for positive identification, will certainly bracket them into the correct fibre group, e.g., scales on the surface of wool and hair fibres (naturally occurring protein fibres), convolutions in cotton fibres and cross-markings in bast and leaf fibres (naturally occurring cellulose fibres). It must be stressed, however, that crimping, bulking, or texturizing processes, chemical treatments, and finishing processes can very easily distort the original form, especially of man-made fibres, and this is an additional complication in fibre identification. The normal light microscope is essential for distinguishing the different types of animal fibres and vegetable fibres; its use for man-made fibres, however, is somewhat limited, and in its place a polarizing microscope will be found to be most useful.

Three types of microscopical preparation may be required for examination of fibres, whole mounts, cross-sections, and casts; each of these will be considered separately. The first examination of a sample of unknown fibres should be by means of a longitudinal view of the fibre in whole mount. The choice of mounting medium is important, as explained in Section 5.4; it is sufficient to state here that the two mountants generally useful for routine work are liquid paraffin and α-bromo-naphthalene[1]. From this preliminary examination it will be possible, in most instances, to say whether or not the sample is a mixture of different fibres (a blend) or of one type only (homogeneous).

It is not within the scope of this book to teach the principles of light microscopy and it is therefore assumed that the microscope is set up correctly and that the analyst has some knowledge and understanding of polarizing microscopy.

4.1.1 Identifying Characteristics

The following features are the first things to look for.

1. Scale margins—indicating animal fibres.
2. Convolutions, lumen, and reversal zones—indicating cotton.
3. Bundles of fibres with or without cross-markings—indicating bast and leaf fibres.
4. Smooth profile without scales or convolutions, with or without striations—indicating man-made fibres or silk.
5. Flat, ribbon-like profile—indicating metallic yarn or split-film fibres.
6. Fibre inclusions, delustring agents, pigments, flame-retardant agents, anti-static agents, etc.

Animal Fibres (Other Than Silk)

To distinguish between the different animal fibres three preparations, (*a*) whole mount, (*b*) scale casts, and (*c*) cross-sections, may be needed. In

---

[1]Man-made fibres show an affinity for α-bromo-naphthalene. Samples should be examined immediately after mounting otherwise a skin–core effect may be observed.

each of these preparations there are identification features which must be looked for.

Thus in (*a*) the features to look for are the pattern formed by the scale margins, regularity of fibre thickness, type of medulla if present, and pigment distribution; in (*b*) note the type of scale pattern and its variation along the fibres; and in (*c*) note the contour of the fibre cross-section, if a medulla is present note the type and contour and whether or not it is concentric with fibre contour, thickness of cuticle, and distribution of pigment. Methods for making scale casts and cross-sections are given in Sections 4.1.4 and 4.1.5 respectively.

Silk

Cultivated silk (*Bombyx mori*) has a smooth, almost featureless appearance when degummed, but when still in the gummed state the filaments are stuck together in pairs and covered with gum (sericin). Tussah (wild) silk has a flat, ribbon-like appearance with fine longitudinal lines. Anaphe silk (a lesser known wild silk) has transverse cross-markings or nodes associated with corresponding increases in fibre diameter, and may be confused with flax ultimates. All three types mentioned here have characteristic cross-sectional shapes. Cultivated silk has triangular-shaped filaments which are stuck together in the gummed state; tussah has long thin wedge-shaped filaments, and anaphe filaments are roughly triangular in cross-section but when gummed together in pairs give the crescent-shape associated with this fibre.

Cotton and Other Seed Hairs

Cotton fibres are recognized by the presence of a lumen and convolutions, i.e., twists in the fibre, however, the unique feature of cotton fibres is the many reversals in the direction of the convolutions, about every 0·4 mm on average, along the length of the fibre. The places at which the spiral direction of the cellulose layers changes are called reversal zones and show as dark cross-markings when examined between crossed polars at the orthogonal position (i.e., parallel to the direction of vibration of the polarizer or analyser). It should also be noted that, because of the spiral structure of the cellulose layers, a cotton fibre will appear bright irrespective of its position between crossed polars. If, however, the cotton has been mercerized the convolutions may not be easily observed because of the swelling effect of the process, but in all instances, careful examination using polarized light will show the reversal zones. The terminations of cotton are tip and root or tensile fractured end.

Akund and kapok are also single-celled seed hairs but are readily distinguished from cotton by their thin cell walls and absence of convolutions.

Bast, Leaf, and Fruit Fibres

Bast, leaf, and fruit fibres are recognized from their general form as bundles of single fibre cells (ultimates). Each ultimate contains a central lumen and cross-markings are sometimes present. Ultimate dimensions are a valuable guide to identification as is the cross-sectional appearance. Characteristic crystals may be present on the surface although these are best seen after ashing. Methods for ashing and preparing ultimates and cross-sections are described in Sections 5.13 and 5.9 respectively.

Man-made Fibres

Man-made fibres can be recognized by their interference colours when viewed between crossed polars at 45° between the extinction positions through a 1st-order Red Plate added to the lens system.

Additionally the birefringence of the fibres can be measured and the cross-sectional shape should be noted. The melting points of thermoplastic fibres are best measured using hot-stage microscopy (see Section 4.1.3).

4.1.2   Interference Microscopy and Birefringence

The majority of textile fibres are optically anisotropic and, therefore, birefringent, that is, the value of the refractive index $n_{||}$, parallel to the axis or along the fibre, is different from that of the refractive index $n_\perp$, perpendicular to the axis or across the fibre. The difference between these two refractive indices $(n_{||} - n_\perp)$ is known as the 'birefringence'. When a birefringent fibre is examined between crossed polars under a polarizing microscope, interference colour bands are generally seen when the fibre is at the 45° position. These interference colours, or the lack of them, can be used to identify fibres, particularly man-made fibres. The introduction of a 1st-order Red Plate (Quartz-sensitive Tint Plate) into the lens system between the crossed polars will generally give specific colours for particular fibres, similar to those described in Table 4.1. The 1st-order Red Plate is particularly useful for fibres of low birefringence. Such fibres emit distinct colours,

TABLE 4.1
Fibre Identification Chart for Polarizing Microscopy[2]*

Fibre	Fibre Parallel to Slow Direction of 1st-order Red Plate	Fibre Perpendicular to Slow Direction of 1st-order Red Plate
Viscose	Pale green, green, dark band, yellow	Yellow, green, dark green, yellow
Acetate	Very dull	
Triacetate	Blue	Orange
	(both dull)	
Polyamide	Yellow-orange, green, red, yellow	Blue-green, red-yellow, green, yellow, or centre pale green or yellow *or* bright green, red, yellow-green, dark grey, orange, centre may be magenta
Polyester	Yellow, green, yellow, or green, yellow, green, yellow	Yellow-orange, green, red, yellow
Acrylic	Yellow or yellow-orange	Blue or purple-blue
Chlorofibre	Turquoise	Yellow

*These are examples of interference colours that may be seen using a 1st-order Red Plate.

---

[2]Many of the man-made fibres may be delustred to a varying extent. Delustring agents will appear in the form of irregular-sized black particles in both polarized and unpolarized light.

e.g., orange in the parallel position and blue in the perpendicular position for acrylic fibres and the reverse order for acetate fibres. Both these fibres appear white in both directions between crossed polars without the 1st-order Red Plate. The colours may vary with fibre thickness and it is recommended that each microscopist should build up a library of known samples. This evidence can be used towards fibre identification. (See Figs 162–167.)

When a man-made fibre is examined between crossed polars and at the 45° position, interference colours are seen. The order of interference at the centre of the fibre depends on (*a*) the type of fibre and (*b*) fibre thickness. For example (without 1st-order Red Plate) cellulose acetate exhibits 1st-order white interference colour even up to 5 dtex. Nylon 6.6, however, exhibits second and third order colours at 3·3 dtex, reaching fifth order colours at 17 dtex, whereas polyester fibres have fourth and fifth order colours at 2·3 dtex and can reach seventh and eighth order colours at 6·7 dtex. If the order of interference is high the 'colours' seen are very pale and polyester fibres can appear white, i.e., colourless, in the diagonal position between crossed polars. It is always advisable when there is doubt to cut the fibre so as to produce a wedge-shaped end (see Diag. 4.1) and with the cut wedge surface uppermost to count the whole orders of interference using monochromatic light.

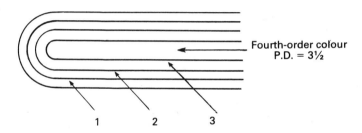

Diag. 4.1 Wedge-shaped fibre end showing fringes

Examples of colours seen for first, second, and third orders of interference:

1st-order colours	Black→grey→white→yellow→bright yellow→brown yellow→red orange→red
Path difference	0–550 nm
2nd-order colours	Deep red→purple→violet→indigo→blue →greenish blue→green→greenish yellow →orange→orange red→dark violet→red
Path difference	551–1100 nm
3rd-order colours	Bluish-violet→indigo→greenish blue →seagreen→green→greenish yellow →carmine red
Path difference	1101–1600 nm.

Acrylic fibres have a low value of birefringence and usually show 1st-order white interference colour, however, these fibres are negatively birefringent. The sign of birefringence can be checked using a 1st-order Red Plate. When the 'slow' direction of the plate (usually indicated by a + sign) is aligned parallel to the fibre axis, with the fibre at the 45° position between crossed polars, the order of interference (colour) will increase for positively birefringent fibres and decrease for negatively birefringent fibres. When the 'slow' of the plate is set across the fibre axis, the converse is true. This test can be carried out only on fibres with low birefringence values, i.e., those exhibiting first order colours, the colours obtained with higher orders of interference are confusing.

It should be mentioned here that the order of interference, i.e., the colour seen at the centre of a fibre when viewed in the diagonal position between crossed polars, is traditionally referred to as retardation. However, it is the distance of displacement in air of the two beams emerging from the birefringent fibre which is actually being measured. This value is strictly called path difference (P.D.), there being a slight difference in value between P.D. and retardation. As it is P.D. that is being measured, it should always be referred to as P.D. and not retardation.

$$\Delta_n \text{ (birefringence)} = \frac{\text{path difference (P.D.) } \mu m}{\text{fibre thickness (t) } \mu m}.$$

The simplest method of estimating P.D. is to cut a wedge-shaped end. Arrange the fibres parallel to one another on a clean hard surface, and using a clean, sharp razor blade cut the fibres, holding the razor blade at a very oblique angle. The cut ends are then mounted in liquid paraffin on a microscope slide and covered with a cover glass. The fibre should be examined with the cut face of the wedge uppermost (this position can be achieved by gently moving the cover glass with a mounted needle). The fibres should be examined between crossed polars in the diagonal position with either a 20× or 40× objective depending on the thickness of the fibre. It is preferable to count the number of fringes in monochromatic light (mercury green, $\lambda = 546 \cdot 1$ nm) (see Diag. 4.1).

P.D. is $n\lambda + \frac{1}{2}\lambda$,

where n = number of whole wavelengths (fringes) of light,
and λ = wavelength of light.

By adding $\frac{1}{2}\lambda$, any error in the estimation is reduced. The thickness of the fibre has to be measured. Cylindrical fibres can be measured with a measuring eyepiece or eyepiece graticule, but irregularly shaped fibres present greater difficulties. The birefringence of the fibre is determined from the formula above.

The P.D. can be accurately determined by using (a) a graduated quartz wedge in conjunction with a Dick Wright eyepiece (see Section 5.4), (b) a rotary compensator, e.g., Berek Compensator (see Section 5.4), or (c), if the microscope is fitted with a graduated rotatable analyser, a Senarmont Compensator for very accurate measurement. It should be stressed that for general fibre identification a fairly close estimation of path difference is all that is required.

Table B3.2 (Appendix B) gives a list of birefringence values for various fibres. These provide a useful guide for fibre identification.

Difficulties may be encountered when examining fibres with a polarizing microscope in the following instances:

(a) where filaments have been texturized, resulting in distortion of cross-sectional shape;

(*b*) filaments and fibres from highly-twisted yarns where fibre shape may have been distorted, particularly if the fabric has been heat-set;

(*c*) deeply-dyed fibres, where the coloration of the fibre may mask the interference colours;

(*d*) heavily delustred fibres, where the delustring agent prevents or restricts the transmission of light, (this also applies to colour pigments); and

(*e*) bicomponent fibres.

Confirmatory tests may be required to complete identification of man-made fibres, such as solubility tests, staining techniques, and melting point determination.

4.1.3 Hot-stage Microscopy

Examination of the effect of heat on fibres is particularly useful for synthetic-polymer fibres, since the natural fibres and natural-polymer man-made fibres do not melt. Melting points are not necessarily conclusive for identification, but they provide useful confirmation. They also provide the easiest method of distinguishing between, for example, the different types of nylon fibres and the components of heterofil fibres. To be of value, fibre melting points must be determined in a standard manner. The Kofler-type hot stage is a simple instrument which provides good thermal contact with the specimen. Although the method described below is based upon the Kofler-type stage, the principles also apply to other hot stages (see Section 5.8). The important point is to ensure good thermal contact between the stage and fibre; a cover glass of convenient size is superior to a microscope slide for this purpose.

Cut the fibre specimen into lengths of 5–10 mm, and spread these evenly on the cover slip. Ideally, all portions should lie in a single layer, but this is not always possible. Cover the specimen with another cover slip and, in the case of the Reichert–Kofler stage, cover this with the glass bridge and close the chamber with the glass plate. Select the correct-range thermometer for the expected melting point and switch on the heater. While the initial warming-up period is progressing, examine the specimen and select an appropriate field. Observe this field thereafter, and note the temperatures at which changes take place, for example,

(*a*)   start of contraction,
(*b*)   finish of contraction,
(*c*)   start of softening,
(*d*)   start of melting, and
(*e*)   finish of melting.

If a fibre mixture is present, the individual different fibres must be located and each observed during the rise in temperature.

The rise in temperature must be constant and uniform, and should not exceed 4 °C per minute until a point within 10 °C of the melting point is reached, when the rate should be dropped to 2 °C per minute.

It is usual in hot-stage analysis to perform an initial sighting run to obtain an indication of the melting point, followed by three determinations under carefully controlled conditions. This, however, may not always be possible due to lack of sample, and thus all determinations should be carried out very carefully.

The temperatures quoted in the literature for fibre melting points often cause misunderstanding, as these can be as much as 20 °C below the true value owing to different methods of determination. For synthetic-polymer fibres the melting point is defined as the temperature at which the fibres comprising the specimen flow easily into their neighbours and no further fibrous form is visible. It will be found that this value agrees with that obtained by differential thermal analysis and similar more sophisticated techniques. (See Table XVI.)

4.1.4 Examination of the Surface of Animal Fibres

It is frequently difficult to see the scale pattern on fibres that are dyed to very dark shades, are densely pigmented, or have a wide medulla. In any of these situations it is necessary to make a cast of the fibre in order to be able to study the scale pattern.

4.1.4.1 Casts Made Without the Use of Heat

Good casts may be made using a 3% solution of gelatin in water to which a crystal of thymol has been added as a preservative. A thin layer of this solution is spread evenly over a microscope slide and the fibres placed in position on the solution. The slide is allowed to dry thoroughly and when the gelatin has hardened the fibres can be lifted away. Alternatively, the slide may be allowed to dry first and then the fibres are laid on the gelatin, a drop of water is added and allowed to run along the fibres. When the slide has been allowed to drain and dry, the fibres can then be removed. This method gives good results with all kinds of fibres.

Other cold methods which may be useful are:
1. cellulose acetate photographic film base and acetone, and
2. nail varnish.

In the first method the fibres are attached at the ends to the film base with cellulose tape and acetone is run along the fibres, when dry the fibres are removed and the film is placed on a slide and examined. In the second method a thin layer of nail varnish is applied to a slide, the fibres are laid on the varnish and when this has dried the fibres are removed.

4.1.4.2 Casts in Thermoplastic Media

The plastics material used is poly(vinyl acetate), m.p. 80 °C, in the form of a 20% solution in benzene: 0·3 ml of the solution spread over about two-thirds of the area of the usual 3 × 1 in. (76 × 25 mm) microscope slide gives a layer of suitable thickness for making a scale impression of a fibre about 22 μm in diameter. The slide is heated to drive off the benzene, leaving a film of poly(vinyl acetate) about 125 μm thick, and this is allowed to cool. A few fibres are placed in the required position on the film, another slide is placed on top, and both are put on a hot plate and weighted to give suitable pressure[3]. A small piece of the solid medium is placed on an additional slide on the hot-plate as a temperature guide. When the medium softens, the heat is turned off and the slides are allowed to cool until the medium becomes hard. The two slides with the medium and fibres between

---

[3]A. B. Wildman. 'The Microscopy of Animal Textile Fibres', W.I.R.A., Leeds, 1954, p. 38, fig. 19.

them may now be separated and the fibres removed. If there is any difficulty in removing the fibres by needles and forceps, the fibres may be dissolved away in dilute caustic soda solution.

A thicker layer of medium must be used for the coarser fibres. This method yields excellent results with kemp and heavily medullated fibres as the weights on the top slide appear to fracture the cortex of these fibres giving impressions of flattened, instead of the usual rounded fibres. Thus a cast of greater area than usual of fibre cuticle is made available for examination.

4.1.4.3 Rolled Impressions

The above methods give casts that are representative of only a portion of the fibre surface. Since scale patterns may vary round the circumference of a fibre, e.g., fur fibres, it is sometimes desirable to prepare impressions of the entire surface. To obtain rolled impressions of the surface, two or three drops of the impression medium are dropped on to a sheet of plate glass, and, by means of the edge of a microscope slide, a very thin film of the medium is spread over an area of about 75×25 mm. A similar film is spread over one side of a previously cleaned slide. On both long edges of the base-plate film are placed spots of wool wax, or material of similar consistency, the distance separating each pair being about 10–20 mm. Single fibres are then mounted in the wax spots so that each fibre is held at a point a short distance from each end. The microscope slide is promptly placed film side down over the fibres and rolled lengthwise, a slight pressure being applied by the thumb. The fibres revolve in the wool wax bearings and if the operation is proceeding successfully a track of impressions becomes visible on both upper and lower films. It is essential that the fibre is free from twist and that the medium is thin and only just tacky immediately before rolling is carried out.

4.1.5 Methods of Cutting Cross-Sections

Cross-sections can provide much useful information for the identification of many fibres, particularly those of animal origin. They are not, however, necessarily a conclusive test for man-made fibres, although they can provide useful confirmatory evidence. A variety of shapes can be given at will to the same man-made fibre during manufacture and various distortions are introduced in further processing such as texturizing. The cross-sections in the figures provide examples of the application of different methods: plate method, Fig. 103; hardy microtome, Fig. 3; hand microtome, gelatine-wax embedding, Fig. 144; mechanical microtome, methacrylate embedding, Fig. 114.

4.1.5.1 Plate Method[4]

This is the simplest and quickest way to cut specimens suitable for identification purposes.

---

[4]J. M. Preston. *J. Text. Inst.*, 1936, **27**, T216.

Apparatus[5] (i) Metal plate 25 S.W.G. (i.e., thickness 0·5 mm in the form of a microscope slide containing holes 0·75 mm in diameter. The plate should be smooth with no protruding ridges round the holes or edges.

(ii) Sharp safety razor blades.

Procedure A tuft of fibres, sufficient to pack the hole, is pulled through by means of a loop of 50 dtex nylon yarn. If insufficient fibre is available, the hole can first be filled by pulling through a bundle of some other readily distinguishable fibre, e.g., delustred cellulose acetate. The protruding tuft of acetate is then opened out in order that the specimen to be cut may be inserted near the centre. The bundle is pulled through a little further so that the specimen and packing comfortably fill the hole.

The projecting tufts of fibres on both sides are trimmed with scissors, and then, with a new safety razor blade, are cut flush with the surface of the plate, the second surface to be cut being retained uppermost. A drop of n-decane or silicone fluid is applied and a cover glass placed on the section. Sections thus produced are examined under the microscope by transmitted light. Because the fibres have a higher refractive index than the surrounding medium, light is transmitted through them by total internal reflection which causes them to appear bright against a dark ground.

The method just described gives excellent results with man-made, wool, and silk fibres, but with cotton, bast, and leaf fibres and delustred or pigmented viscose, a modification can be used to advantage. Various substances are used to obtain contrast between the fibres and the background, such as black oil paint[6] or certain solutions[7]. The black oil paint is carefully rubbed over the surface of the section filling the inter-fibre spaces and the excess wiped off using a clean cloth or tissue moistened with n-decane. This method can be used for the majority of fibres.

The use of solutions is of greater benefit in the case of man-made fibres and the solutions used will depend on the amount of pigmentation or delustrant present in the fibres. In general terms, a low pigmentation level requires a low refractive index solution to achieve a suitable degree of contrast. For bright, clear, or lightly-dyed fibres suitable solutions are n-decane (n 1·41), dibutyl phthalate (n 1·49), or Cargille Index of Refraction solution 1·480. Dull or heavily-pigmented fibres require a solution of higher refractive index such as 1-bromonaphthalene (n 1·658) or Cargille Index of Refraction solution 1·690.

Sections are prepared using the plate method and a drop of the appropriate solution is applied to the top surface of the section followed by a cover slip (10 mm diameter). The appearance of the section when examined microscopically will depend upon the fibres and upon the solution used. Bright fibres will appear white against a darker background and dull fibres will be black against a white background. In some cases it is difficult to achieve a significant degree of contrast with bright man-made fibres, but with the addition of a small amount of suitable dyestuff, e.g., Sudan Red II or IV, to the solution and the use of a blue-green filter, such as the Ilford Micro 2 or Kodak Wratten 75, a very good degree of contrast can be obtained.

---

[5]Kit obtainable from Shirley Developments Ltd.

[6]Ivory Black Oil Colour, obtainable from Winsor and Newton Ltd, London.

[7]J. E. Ford and S. C. Simmens. *J. Text. Inst.*, 1959, **50**, P148.

4.1.5.2 Hardy Microtome[8]

Apparatus This microtome consists of two metal plates whose edges are held in alignment by grooved pieces. One plate is slotted down the centre to receive the fibre bundle; the other is made with a tongue for forcing the fibres down to the base of this slot. The two plates fit together, tongue into slot, to form a single unit having the size and shape of a microscope slide.

Procedure Fibres combed parallel for sectioning are inserted into the slot before the plates are pushed together. After the protruding ends of the fibres have been cut off flush with the metal on each side, a plunger, which fits snugly in the slot, is mounted upon the slotted plate by means of a small screw. The plunger is moved by a fine screw and this is used to push through the bundle of fibres by a few microns before each cut. To prevent the fibres separating after cutting, the projecting bundle of fibres is coated with quick-drying nitrocellulose lacquer (consisting of 3% nitrocellulose in 50–50 absolute alcohol–diethyl ether). Nitrocellulose is obtainable in the form of celloidin shreds. The viscosity of the solution may vary owing to evaporation of the solvent, in which case it can be thinned down by adding alcohol–ether. A spot of lacquer is applied by means of a glass rod or a fine watercolour paint brush to the exposed tuft of fibres. The spot of lacquer is of a suitable size if it extends round the slit with a rim equal to the width of the slot. This is found to prevent the excessive rolling of the section that occurs when only the projecting ends of the fibres are coated. The thickness of the dried film of lacquer should be such as to give sufficient mechanical strength for handling. The actual cut is made with a safety razor blade by moving the two hands at right angles, thus making a slicing cut. The small group of cut and embedded fibres is transferred to a glass microscope slide, mounted, and examined in the usual way. Full instructions are provided with the instrument.

4.1.5.3 Thin Cross-sections

This method is suitable for obtaining thin sections of fibres and yarns where the area of cross-section to be cut is no more than 2–3 mm². Cutting sections of bigger areas causes problems such as uneven section thickness and also more damage to the knife.

Apparatus A mechanical microtome of the rocking, rotary, or sliding (sledge) type is required, capable of advancing the embedded section by increments of one or more microns. In these microtomes the knife is either held rigidly or guided mechanically, but in a fourth type, the so-called hand microtome, the knife is guided by hand.

Procedure Samples must be held in a suitable former from which they can be easily removed after embedding. Beem-type capsules are recommended; these are made of polythene and are available in a variety of shapes and sizes, all having hinged, airtight caps. The interiors of the capsules are shaped to give pre-shaped blocks requiring a minimum of trimming before sectioning. The fibre or yarn samples are threaded through holes, made centrally both in the cap and tip of the capsules, by

[8]J. T. Hardy. U.S. Dept. of Agric. Circ. 378, 1935; T. L. W. Bailey and M. L. Rollins. *Text. Res. J.*, 1945, **15**, 1; A. B. Wildman. 'The Microscopy of Animal Textile Fibres', W.I.R.A., Leeds, 1954.

means of a thin wire loop or sewing needle. The bottom end of the sample is held firmly in place with cellulose tape which also seals the hole in the capsule tip. The capsule is then filled with the embedding resin and the cap closed. The sample is tensioned so that it is parallel to the walls of the capsule and held in place with cellulose tape, which again seals the hole in the cap. Once the resin has set, the capsules are removed from the blocks either by slitting with a razor blade or by using a commercial capsule splitter.

An alternative method is for the thoroughly mixed ingredients to be placed in a vacuum oven at 50–60 °C under vacuum for a few minutes. Air is pulled out of the mixture and the viscosity of the mixture is lowered. The vacuum is released and the mixture removed from the oven. The capsules are three-quarters filled and replaced under vacuum to remove air from between the fibres. They are then removed from the oven, topped up with Araldite, and closed. The filaments are gently pulled straight in the capsules and the free end stuck down with cellulose tape. The capsules are replaced in the oven at 50–60 °C for 24 hours to allow the Araldite to 'cure'. No vacuum is required for this final stage.

Two types of synthetic polymers are recommended, Araldite[9] and methacrylate resins[10], but other synthetic-polymer resins are available. These resins can be obtained directly from the manufacturers or from suppliers of microscopical equipment. The embedding recipes given both by suppliers or in books on microscopy are designed generally for section cutting with glass or diamond knives for electron microscopy. It is recommended that slightly more plasticizer is used in Araldite resins and that the 'hardness' of methacrylate resins is altered by varying the proportions of methyl to butyl methacrylate. The two recipes given here are a basis for embedding and can be adapted to suit individual requirements.

Araldite Epoxy Resin[9]

Araldite Casting Resin CY212	20	ml
Hardener HY964	20	ml
Dibutyl phthalate (plasticizer)	3	ml
Accelerator DY 064	0·6	ml

Mix at 50 °C for 30 min, stirring vigorously, then allow to stand for 15 min at 50 °C to exclude air bubbles. Allow the blocks to harden at 60 °C for a period of between 24 to 48 hours.

Methacrylate Resins[10]

n-Butyl methacrylate	90	ml
Methyl methacrylate	10	ml
2, 4-Dichloro-benzoyl peroxide	2·1	g

Mix all ingredients thoroughly, allow air bubbles to escape before pouring into the Beem-type capsules. Allow resin to set overnight at 60 °C.

Note: 2,4-Dichloro-benzoyl peroxide is in paste form to prevent explosion. Care should be taken to keep this paste out of eyes and off hands.

An alternative method which is highly suitable for sectioning animal fibres is a double embedding technique where a 'candle' of fibres is made in a solution of poly(isobutyl methacrylate) in toluene. Subsequently portions of the 'candle' are embedded in wax, e.g., of the Estax or Paramat type. A

---

[9] Obtainable from Ciba-Geigy Ltd.

[10]Obtainable from ICI (Plastics Division).

few fibres are tied together at intervals of approximately 5 mm using a fine thread. This bundle is then suspended from a suitable stand and sufficient weight added to the lower end to keep the fibres straight. The fibres are coated with enough layers of the methacrylate solution to hold them firmly; when this is hard the 'candle' is cut into pieces retaining the portions between the threads. Each portion is then embedded in wax. Finally the blocks are trimmed with a razor blade to a size suitable for mounting on the specimen holder of the microtome.

The sections are mounted on slides previously coated with egg albumen solution, covered with a drop of water, and allowed to relax by gently warming the slide. The water is dried off and the wax removed in xylol; afterwards the sections are mounted in some suitable mountant, e.g., liquid paraffin (n 1·47).

4.1.5.4 Grinding Method

The former methods are adequate for the preparation of fibre cross-sections and require various levels of skill and experience. However, these methods are not versatile enough to accommodate large specimens covering several square millimetres or where it is important to preserve the spatial arrangements of fibres. A fairly simple method of obtaining cross-sections is the grinding method[11]. This technique requires relatively little skill in specimen preparation, large flat sections can be obtained, and equipment costs are low.

Apparatus[12] The prepared blocks are ground on metallurgical grinding equipment, a suitable instrument being the Lunn Minor[13] which takes four grades of silicon carbide paper. These papers are washed by a continuous flow of water to remove debris resulting from grinding and this also prevents heating up of the sample. The four grades of paper, numbers 220, 320, 400, and 600 have particle sizes of 70, 35, 25, and 15 μm respectively. The two coarser grades are used to grind away most of the surface resin and material and the finer grades reduce the depth of the scorelines in the specimen surface.

Procedure The specimens must be held in a former that can be easily stripped from the block after the resin has set or that will not hamper grinding. Fibres and yarns can be held in Beem-type capsules as recommended in Section 4.1.5.3 on thin cross-sections. Fabrics are best mounted across a cardboard window which is then held upright in a holder fashioned from aluminium foil (kitchen baking-foil). The edges of the foil must be completely sealed to prevent the resin leaking away and aluminium holders must be held upright while the resin sets. A better, but more costly method, is to make a silicone-rubber mould with rectangular slots to hold the cardboard sample holders (see Diag. 4.2). The fibres, yarns, or fabrics are again secured across cardboard windows using Araldite adhesive from the twin-tube pack and are inserted into the slots in the rubber mould. The casting resin is poured in and allowed to set. The hardened blocks are gently eased out of the mould (there is no adhesion between rubber and resin) and are

[11]B. Lomas and S. C. Simmens. *J. Microsc.*, 1970, **92** (1), pp. 37–45.

[12]Kit available from Shirley Developments Ltd.

[13]Obtainable from Vickers Instruments Ltd, Haxby Road, York.

ready for grinding when surplus resin has been trimmed off, preferably with a small hacksaw. The silicone rubber recommended is Silcoset RVT 101 with Silcoset curing agent A[14,15], this rubber has good tensile and tearing strength coupled with adequate extension properties. The advantage of this method is that the rubber mould can be made to suit individual requirements.

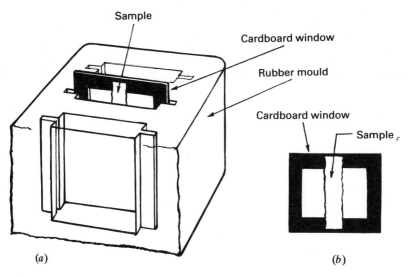

Diag. 4.2 (*a*) Cardboard windows inserted in rubber mould.
(*b*) Cardboard window with fabric specimen attached.

Yarns and fabrics can also be embedded in resin in thin-walled polythene tubing of about 15–20 mm internal diameter. One end of the sample is placed in a slit cut into a rubber bung. This bung serves two purposes, (*a*) to seal the bottom end of the polythene tube, and (*b*) to securely hold the end of the sample. The sample is held in the vertical position by placing its opposite end over a wire positioned across the open top end of the tube. Both wire and sample are held firmly in position by cellulose tape (see Diag. 4.3) while the resin sets. It is advisable to remove the tubing before grinding the sample. More rigid fibres, e.g., bast fibres, can be embedded by placing a number of fibres into a blind hole drilled into a short length of poly(methyl methacrylate) rod. An aluminium foil container is formed round the top end of the rod containing the fibres and held in place and sealed with cellulose tape (see Diag. 4.4). The aluminium former is filled with resin and then placed in a vacuum. On release of the vacuum, the resin fills the hole, and the block is then left for the resin to set. The samples are ground without removing the rod and the resulting sections show the variation in appearance of bast and leaf fibres. Vacuum impregnation can also be used on samples that contain many air spaces, e.g., needle-punched nonwoven fabrics, where air bubbles could spoil the quality of the final section.

---

[14]Obtainable from ICI (Organics Division).
[15]Obtainable from Ciba-Geigy Ltd.

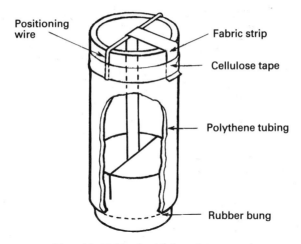

Diag. 4.3. Holder for fabric and yarn samples.

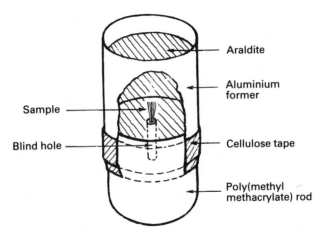

Diag. 4.4. Holder for rigid fibre samples.

The prepared specimens are embedded in a resin which is hard enough and brittle enough to withstand grinding. Araldite Casting Resin CY212 and Hardener HY956[15] are mixed thoroughly in the proportion 5:1 respectively, the resin is left standing to allow time for air bubbles to escape, and is then poured into the formers and allowed to set at room temperature for 24 hours. This resin is recommended for its very low shrinkage and compression properties, as well as its low viscosity.

When the resin has set, the former is removed, any surplus resin trimmed off, and one face of the block ground flat on the silicon carbide papers so that this ground face is perpendicular to the specimen length. A thin disc, about 2–4 mm thick, is cut from the block (a small hacksaw is recommended) and the ground face is stuck onto a microscope slide with a thin layer of Araldite (twin-tube pack) which is allowed to set at a temperature of 60–70 °C for about 30 min. The upper face of the disc is now ground by the same method used for the lower face. Difficulties arise in holding the microscope slide and grinding parallel to the lower face. It is recommended that a small holder be made with a recessed plate, shallower than the thickness of a microscope slide, into which the prepared slide can be secured with double-sided adhesive tape[16] whilst grinding. A recommended section thickness is 20–30 μm.

The refractive index of Araldite resins is similar to that of most fibres so, in order to increase contrast between resin and undyed material, the sections can be stained in a hot solution of either Durazol Brilliant Blue or hot Shirlastain A for 10 minutes. When stained, the sections are thoroughly washed and dried and then mounted in tritolyl phosphate (tricresyl phosphate). Greater contrast can be obtained by using colour filters. The effect of staining is to colour the resin leaving the fibre cross-sections unstained; the exceptions being those fibres which have an affinity for acid dyes. It is advisable therefore not to stain wool, silk, or ligno-cellulose fibre sections in Shirlastain A.

Samples may be examined after the first ground face has been obtained. This face is mounted in liquid paraffin and examined under a stereo microscope using diffused reflected light. This type of examination is particularly useful with fabric samples where the interlacing of the yarns can be studied.

4.1.5.5   Cross-sections of Glass Fibres

Cross-sections of glass fibres cannot be cut by conventional means, but plate sections may be made as follows: the fibres are threaded through a hole in a metal plate and embedded in position with shellac, so that the shellac penetrates throughout the bundle. The shellac is ground flush with the plate on both faces, and polished until smooth. The resulting plate section is examined in the usual way.

4.2   Identification of Homogeneous Fibre Specimens

4.2.1   Preliminary Sorting Tests by Heating[17] (Table I)

TABLE I

Apply each of the three sorting tests given below and classify the fibre according to its behaviour in the tests. Then proceed as instructed in the last column of the table.

Test 1 Take a small tuft of the fibres, hold in forceps if necessary, and bring it slowly up to, but not into, a small non-luminous flame. Observe whether the fibres shrink or melt.

Test 2 Place a small tuft of fibres close to a crystal of potassium nitrate on a metal plate that can be heated. Raise the temperature of the plate until the crystal melts, which is at a temperature of 337–339 °C. Observe whether the fibres char or melt, the fibres having been prodded with a pointed needle.

Test 3 Bring a small tuft of fibres into a small non-luminous flame. Observe whether the fibres burn, and note the smell of the vapour produced.

---

[16]Available from William Hayes, Bankfield Works, Boothtown, Halifax.

[17]See also Section 4.2.6, Preliminary Sorting Tests by Solubility.

TABLE I

1 Approach to Flame	Behaviour of the Specimen		Fibre Type	Further Procedure
	2 On Hot Plate	3 In Flame		
Does not shrink from flame or melt	Chars below 337 °C. Does not melt	Burns with irregular spurting flame, leaving a black, inflated, easily-powdered residue, and emits a smell like that of burnt hair[1]	PROTEIN	Proceed as in Tables VII or VII (a)
		Burns readily, emitting a smell like that of burnt paper, leaving a small amount of ash (or sometimes emitting a distinct fishy odour, leaving a dark, skeletal residue)	CELLULOSE (or RESIN TREATED CEL-LULOSE)	Proceed as in Table VIII
		Burns slowly, extinguishing when flame removed. Burning may be accompanied by evolution of fumes having a pungent odour. Carbonized skeletal residue.	FR VISCOSE OR CEL-LULOSE WITH FR FINISH	Proceed as in Table VIII
		Burns readily, extinguishing at once on removal from flame, leaving an incandescent residue	CALCIUM ALGINATE	Proceed as in Table IX
	Does not char or melt[2]	Melts to a clear hard bead	GLASS	Proceed as in Table X
		Glows, but retains its original form	ASBESTOS[3]	Proceed as in Table X
Shrinks or melts to a bead	Melts below 337 °C	Burns and drips from the flame	THERMOPLASTIC AT LOW TEMPERATURE	Proceed as in Table II
	Does not melt up to 337 °C	Burns and drips from the flame or chars	ARAMID	See Table V (a)
		Does not burn	TEFLON	See Confirmatory Test

[1]Tin-weighted silk shows a skeletal residue that glows in a flame.
[2]In the case of glass, smoke may be emitted by dressings and there may be discoloration.
[3]It should be remembered that asbestos is commonly blended with cotton or some other fibre.

L

Confirmatory Test for Teflon
The refractive index of Teflon is much lower than that of other fibres.
Examine the specimen in acetone under the microscope, preferably using
light polarized perpendicular to the fibre axis, and apply the Becke rule
(see Section 5.4). A refractive index below the 1·359 of acetone identifies
Teflon fibres.

4.2.2   Fibres Thermoplastic at Low Temperature (Tables II–VI)

TABLE II
1.   Test a specimen for the presence of nitrogen.
     Apply the Soda-lime test (see Section 5.17.11).
2.   Test a specimen for the presence of chlorine.
     Apply the Beilstein test (see Section 5.17.2).

The fibres that are thermoplastic at a low temperature fall into one of four
groups.
   Chlorine and nitrogen absent. Proceed as in Table III.
   Chlorine present, nitrogen absent. Proceed as in Table IV.
   Chlorine absent, nitrogen present. Proceed as in Table V.
   Both chlorine and nitrogen present. Proceed as in Table VI.

4.2.2.1 Chlorine and Nitrogen Absent

Cellulose diacetate, cellulose triacetate, poly(vinyl alcohol), polyolefin,
polyester.

TABLE III
Treat the specimen in a test-tube with the following reagents, at room
temperature unless stated otherwise, using a fresh specimen for each test.
Continue with the reagents in succession, in the order given, until one is
found in which the specimen dissolves.

First Reagent of the Series in which the Specimen Dissolves	The Specimen is
70% v/v aqueous acetone	**CELLULOSE DIACETATE**
Glacial acetic acid	**CELLULOSE TRIACETATE**
5N hydrochloric acid at 65 °C	**POLY(VINYL ALCOHOL)** (see Table III (*a*))
Commercial xylene at the boil	**POLYOLEFIN** (see Table III (*b*))
Insoluble in all these solvents	**POLYESTER** (see Table III (*c*))

Confirmatory Test for Cellulose Diacetate and Cellulose Triacetate
To distinguish further between cellulose diacetate and cellulose triacetate,
boil the fibres for 20 min in 2% sodium carbonate solution, wash in water,
then dye at the boil for 5 min in a 1% solution of a direct dye, e.g., Direct
Fast Black B (Colour Index Direct Black 51) and wash them in cold water.
Cellulose triacetate is substantially undyed; the cellulose diacetate is

heavily dyed. Partially saponified diacetate or triacetate will not be completely soluble in 70% acetone and glacial acetic acid, respectively (see p. 29). This may lead to confusion with other fibres in this group. To detect whether any unsaponified material has dissolved, water should be added to the test-tubes in which the first two solvents have been used. A precipitate will indicate the presence of diacetate or triacetate.

Examination of Vinylal Fibres

Table III (*a*)
Treat the specimen with 5N hydrochloric acid at room temperature.

Behaviour of Specimen	The Specimen is
Dissolves	FORMALDEHYDE-HARDENED
Does not dissolve	BENZALDEHYDE-HARDENED

Confirmatory Tests for Vinylal Fibres
Dissolve the fibres in hot 35% sulphuric acid. Formaldehyde-hardened fibres (normal-tenacity, e.g., 177 mN/tex) give a positive test for formaldehyde (see Section 5.2.2), but benzaldehyde-hardened fibres and high-tenacity fibres (e.g., 442 mN/tex) give a negative result.

Both types of fibre, after dissolving in sulphuric acid, develop a deep blue colour with 0·1% iodine solution.

Benzaldehyde-hardened fibres dissolve in 5N hydrochloric acid at 65 °C.

Examination of Polyolefin Fibres

Aberclare, Courlene X3, Drylene, Fibrite, Gymlene, Herculon, Marvess, Meraklon, Neofil, Pylen, Spunstron, Tritor.

TABLE III (*b*)
Polyolefins are the only fibres, except specially produced hollow types, whose density is below 1·00, hence polyolefin fibres float on water. For this observation the water should contain a wetting agent (e.g., 1 g/l Lissapol N, ICI; Nonidet SH30, Shell Chemicals (UK) Ltd).

Determine the melting point of the fibre (see Section 5.8) and the density (see Section 5.7 or p. 39).

Melting Point (°C)	Density (g/cm³)	The Specimen is
133	0·95	POLYETHYLENE e.g., Courlene X3
160–165	0·90	POLYPROPYLENE e.g., Aberclare, Fibrite, Gymlene, Meraklon, Neofil

Differentiation of Polypropylene Fibres
Dyeing tests carried out in the following order can be used to distinguish between dyeable polypropylene fibres. Approximately 0·05 g fibre in 10 ml of dye solution or dispersion is used for each test which can only be satisfactorily applied to undyed samples.

1. Place the fibre sample in a 0·1% solution of Polypropylene Scarlet RBM (Allied Chemical Corporation), raise to the boil and boil gently for 2 min. Wash the pattern and dry. Nickel-modified fibre (e.g., Herculon Type 40) is dyed to a medium-depth scarlet shade. If nickel is not present then a yellow shade is produced.

2. Immerse the fibre in a 0·1% solution of Lissamine Red B (C.I. Acid Red 37) containing 2% acetic acid. Raise to the boil and boil for 2 min. Wash thoroughly and dry. A heavy bluish-red shade is produced on acid-dyeable fibres. Disperse-dyeable polypropylene (e.g., Polycrest SDR-1) is also stained since a similar type of polymer additive is incorporated in both classes of fibre. The shade produced is, however, much weaker.

It is not possible to distinguish between acid and disperse modified fibres, by the use of disperse dyes because of the close similarity in their dyeing behaviour, but dyeing at the boil for 2 min in a 0·1% Durazol Red 2B (C.I. Direct Red 81) solution produces a medium-depth dull bluish-red shade on acid-dyeable yarn (e.g., Meraklon DR) whereas disperse modified fibre is only very slightly stained.

Modified polypropylene from other manufacturers may possess different staining properties and it is advisable to carry out the above tests on known samples for comparison and future reference.

Examination of Polyester Fibres

Note: The thermal contraction of samples of polyester fibres is dependent on the thermal history of the individual specimen (e.g., temperature of heat setting, finishing method, etc.), and is not meaningful for identification purposes.

TABLE III (c)

1. Treat a fresh specimen with Shirlastain E (see Section 5.17.7).
2. Determine the melting point (see Section 5.8) and/or the fibre density (see Section 5.7).

Shirlastain E	Density (g/cm³)	Melting Point (°C)	The Fibre is
Pink	—	—	**BASIC-DYEABLE POLYESTER** e.g., Dacron 62, 64, 65, and 89, Diolen 41
Unstained	1·23	290–295	**PCDT POLYESTER**[1] e.g., Kodel 200
	1·38	250–260	**PET POLYESTER**[2] e.g., Crimplene, Terylene

[1]Poly(1, 4-cyclohexylene-dimethylene terephthalate).
[2]Poly(ethylene terephthalate).

Confirmatory Test for Polyester Fibres
Place a specimen of the fibre in approximately 10 ml of *o*-phosphoric acid (analytical grade sp. gr. 1·75) in a boiling tube and bring rapidly to the boil

(215 °C). Continue boiling for not more than 1 min. Polyester fibres remain visually unchanged. With a few exceptions, notably glass, PTFE, and asbestos fibres, all other fibres either dissolve with discoloration, melt, or shrivel to gelatinous brown lumps.

Alternative Differentiation between Kodel 200 and PET Polyester

On boiling for 10 min in 10% hydrazine in n-butanol, PET polyesters dissolve (precipitate on cooling), but Kodel 200 is insoluble (no precipitate on cooling).

4.2.2.2   Chlorine Present, Nitrogen Absent

Poly(vinyl chloride), chlorinated poly(vinyl chloride), syndiotactic poly(vinyl chloride), copolymers of vinylidene chloride and vinyl chloride, and of vinyl chloride and vinyl acetate.

TABLE IV

Treat the specimen in a test-tube with the following solvents, at room temperature unless stated otherwise, using a fresh specimen for each test. Continue with the reagents in succession, in the order given, until one is found in which the specimen dissolves.

First Reagent of the Series in which the Specimen Dissolves	The Specimen is
Commercial xylene	CHLORINATED POLY(VINYL CHLORIDE) e.g., Piviacid
Chloroform	COPOLYMER OF VINYL CHLORIDE AND VINYL ACETATE e.g., Vinyon HH
Tetrahydrofuran	POLY(VINYL CHLORIDE) e.g., Rhovyl, Fibravyl, Thermovyl
Commercial xylene at the boil	COPOLYMER OF VINYLIDENE CHLORIDE AND VINYL CHLORIDE e.g., Saran, Velan
Boiling form–dimethylamide (DMF)	SYNDIOTACTIC POLY(VINYL CHLORIDE) e.g., Leavil

Confirmatory Tests for Chlorofibres

Copolymers of vinylidene chloride and vinyl chloride when immersed in morpholine slowly darken and the morpholine becomes nearly black in colour.

Chlorinated poly(vinyl chloride) when immersed in morpholine dissolves and the solution becomes reddish-brown.

Poly(vinyl chloride) and vinyl chloride–vinyl acetate copolymers are not affected by morpholine.

4.2.2.3   Chlorine Absent, Nitrogen Present

Nylons, acrylics, polyurethane elastomers, rubber.

TABLE V

Examine the general nature and extensibility of the specimen.

Behaviour of Specimen	The Specimen is
Normal fibre-like properties	**NYLON, ARAMID or ACRYLIC** (see Table V (*a*))
Rubber-like elasticity, extensibility at least 400%	**POLYURETHANE or RUBBER** (see Table V (*d*))

Distinction between Nylon and Acrylic Fibres

TABLE V (*a*)

Treat the specimen in a test-tube with the following reagents in the order given, using a fresh specimen for each test.

m-Cresol at Room Temperature	Boiling Form- dimethylamide	The Specimen is
Dissolves		**NYLON 6, 6.6, or 11** (see Table V (*b*))
Does not dissolve	Dissolves	**ACRYLIC** (see Table V (*c*))
	Does not dissolve	**ARAMID**

Examination of Nylon and Aramid Fibres

Nylon 6, nylon 6.6, bicomponent nylons (sheath–core or side-by-side), nylon 11, Qiana, Nomex, Kevlar.

TABLE V (*b*)

Carry out the following tests, at room temperature, in the order in which they appear from left to right, and observe whether the fibre dissolves. (See also Section 5.14.)

TABLE V (*b*)

Hydrochloric Acid 4·4N (see Section 5.17.6)	Hydrochloric Acid 5N (see Section 5.17.6)	Calcium Chloride–Formic Acid (see Section 5.17.3)	*m*–Cresol	Cross-sectional Shape	Fibre Type
					Nylons
Soluble				Round	NYLON 6
Surface soluble				Round (core–sheath structure may be visible)	BICOMPONENT NYLON (CORE-SHEATH)[1]
Insoluble	Soluble (5 min)			Round or trilobal	NYLON 6.6
	Insoluble (5 min)	Soluble		Round (bilateral structure may be visible)	BICOMPONENT NYLON (SIDE-BY-SIDE)[2] e.g., Cantrece
		Insoluble	Soluble	Round	NYLON 11 e.g., Rilsan
				Trilobal	QIANA
			Insoluble	Peanut	NOMEX
				Round	KEVLAR

(Fibre Type column also carries the group heading *Aramids*[3] above NOMEX and KEVLAR.)

[1] With bicomponent fibres of nylon 6.6 surrounded by nylon 6, the sheath can be seen to dissolve in 4·4N HCl under the microscope.

[2] Cantrece is nylon 6.6 spun side-by-side with a copolyamide.

[3] Aramids should have been detected by the procedure given in Table I. Nomex burns and drips from the flame, Kevlar chars.

Examination of Acrylic Fibres

TABLE V (c)

Perform the following tests in the order in which they appear from left to right. It is convenient to perform the Meldrum staining test and the first solubility test with nitromethane, simultaneously in the same water-bath at 90 °C.

Meldrum's Stain At 90 °C for 5 min (see Sections 5.15, 5.16)	Nitromethane (highly inflammable)				Cross-sectional Shape (see Section 4.1.5)	Sevron Orange Boil for 5 min[1] (see Sections 5.15, 5.16)	Form–dimethylamide Temperature in °C required to dissolve in 2–3 min[1] (see Section 5.16)	Fibre Type
	At 90 °C for 5 min	At 60 °C for 3 min	At 65 °C for 3 min	At 75 °C for 3 min				
1	2	3	4	5	6	7	8	*Staple Fibres*
Pale yellow	Insoluble						35	COURTELLE
Pale apple green							80	ZEFRAN
Deep green	Soluble				Peanut			ORLON 44
					Round			EXLAN S
	Insoluble						55	ACSA-N
Red shades (including scarlet, red, orange, pink. Disregard depth of shade at this stage)	Soluble	Soluble						CRESLAN 61 CRILENKA
		Insoluble	Soluble			Orange		EXLAN L
						Scarlet		CASHMILON
			Insoluble	Soluble		Orange	25	CRESLAN 67A
							35	CRESLAN 61B

Table V (c)—(continued)

1	2	3	4	5	6	7	8	
Red shades including scarlet, red, orange, pink. Disregard depth of shade at this stage	Soluble	Insoluble	Insoluble	Soluble	Peanut	Scarlet		EUROACRIL EXLAN Dkp
					Peanut			ORLON 42
					Round	Pink		CRESLAN 68B
					Round	Orange		CRESLAN 68CS
				Insoluble			35	VONNEL 17
					Bean			ACRILAN 16
								ACRILAN 57
							40	ACRILAN 70
								ACRILAN 71
								ACSA–16
	Insoluble				Peanut		35	DOLAN
								REDON
							40	DRALON
					Bean	Red		TORAYLON ACRIBEL
					Irregular bean	Pink	45	CRYLOR TYPE 20
					Round or oval		50	CRYLOR TYPE 50
					Round	Orange		BESLON[2]
					Lobed (bicomponent)	Magenta		ORLON 21

1	2	3	4	5	6	7	8
							Filament Yarns
Red shades					Bean, peanut, or pear	Scarlet	CRYLOR 50
					Peanut	Pink	DRALON 40

[1] Fine-denier fibre dyes paler than coarse. Pigmented fibre dyes paler than bright.
[2] Beslon disintegrates in nitromethane at 90°C.

Examination of Polyurethane Elastomers and Rubber

Lycra, Dorlaston, Enkaswing, Glospan, Sarlane, Spanzelle, Lustreen, Rubber.

Elastomeric yarns are continuous filament and may be 'bare' or wrapped spirally with a non-stretch yarn, e.g., cotton, viscose, or nylon; or core-spun with a sheath of staple fibre round the elastomeric core.

TABLE V (*d*)

Separate the composite yarns manually for examination of the elastomer. Ignite the sample. Rubber emits a distinctive smell, usually sufficient for its identification. Polyurethanes burn with a luminous flame but are not distinguishable by smell. A further means of differentiation is the solubility of the elastomer in cold conc. nitric acid; polyurethanes are soluble within 15 minutes, rubber takes several hours to dissolve.

The different types of polyurethane elastomer can, to some extent, be distinguished by their solubilities in form–dimethylamide and caustic soda, the former indicating the degree of chain branching or cross-linking of the polymer and the latter the presence of ester or ether linkages. Nevertheless, it should be emphasized that yarn producers do sometimes change polymer systems without necessarily changing trade names, so the details in the table below should be used only for general guidance.

Solubility in Boiling Form-dimethylamide	Solubility in Boiling 10% NaOH for 30 min	The Specimen is
Soluble (<10 min)	Insoluble	LYCRA 126 LYCRA 124
	Disintegrates	DORLASTON SARLANE ENKASWING
Partly soluble (approx. 15 min)		GLOSPAN
Insoluble	Soluble	SPANZELLE, LUSTREEN

4.2.2.4 Chlorine and Nitrogen Present

Examination of Modacrylic Fibres

Dynel, Kanecaron, Teklan, Verel, Crylor, SEF Modacrylic.

TABLE VI

Observe the behaviour of the specimen in the following tests.

Butyrolactone for 2 min	Nitromethane for 2 min	Acetone at 35 °C for 5 min	Cross-sectional Shape	Fibre Type
Soluble	Soluble	Soluble		DYNEL 150
				DYNEL 180 KANECARON
				DYNEL 197
		Insoluble		TEKLAN
	Insoluble			SEF MODACRYLIC
				VEREL, types A and F
Insoluble				CRYLOR PCM

Confirmatory Test for Modacrylics
Modacrylic fibres can be distinguished by their infrared spectra, which are highly characteristic.

4.2.3   Fibres Non-thermoplastic at Low Temperature (Tables VII–IX)

4.2.3.1   *Protein Fibres*

Protein Fibres: Identification by Microscopical Appearance

Wool, llama fibres, camel hair, mohair, cashmere, common goat hair, rabbit and hare fur, horse hair, cow hair, silk, regenerated protein fibres.

TABLE VII

Examine the fibre microscopically in longitudinal view in liquid paraffin, and in cross-section.

TABLE VII

Longitudinal View	Cross-sectional View	The Specimen is
Irregular diameter and prominent scale margins[1]. Medulla present in some medium and coarse fibres; may be fragmental, interrupted, or continuous.	Oval to circular, variable in diameter. Medulla, if present, is concentric and variable in size.	**WOOL.** Figs 1–15
Regular diameter and smooth profile. Scales shallow.	Fine fibres—oval to circular, coarse fibres—oval, circular, triangular, or polygonal. Fairly thick cuticular layer of scales. Pigment is sparse to very dense, some in large aggregates. Medulla is usually present and a bi- or multipartite medulla is frequent in coarse fibres.	**LLAMA FIBRES** Figs 24, 25
Regular diameter and smooth profile. Scales shallow. Wide range of fibre diameter. Pigment—frequently streaky. Medulla—none or fragmental in fine fibres, interrupted or continuous in coarse fibres.	Oval to circular, fibre diameters variable in size in the fleece. (Fine fibres are used in apparel fabrics, coarse fibres in mechanical cloths.) Pigment is sparse to dense. Medullae in coarse fibres small in relation in total area.	**CAMEL HAIR** Figs 22, 23
Regular diameter and smooth profile; scales very shallow. small vacuoles appearing black in some fibres.	Circular to oval. Pigment—none or very sparse. Medulla—occasionally fragmental or continuous in the coarser fibres.	**MOHAIR** Figs 16, 17
Fairly regular in diameter with prominent scale margins. Pigment—none to dense. Medulla—interrupted or continuous in coarse fibres.	Fine fibres—almost circular. Coarse fibres—oval to circular; some flattened, often medullated. (Coarse fibres are not found in yarns or fabrics.)	**CASHMERE** Figs 18, 19

[1]Wools that have received certain chemical treatments may have scales that are much less prominent than those of untreated wools. Mechanically damaged wools (e.g., re-used wools) may, in addition, be fibrillated at places along the fibre length.

TABLE VII—*continued*

Longitudinal View	Cross-sectional View	The Specimen is
Fine fibres—irregular diameter with prominent scales. Coarse fibres—fairly regular in diameter and fairly prominent scales. Pigment—none to very dense. Medulla—none in fine fibres. Coarse fibres—none, fragmental, interrupted, or continuous.	Fine fibres are almost circular. Coarse fibres may be almost circular, irregular, or flattened; when medullated the medulla is frequently of the wide lattice type.	COMMON GOAT HAIR Figs 20, 21
Fairly regular diameter with prominent scale margins. Medulla—either uni- or multi-serial ladder type, wide in relation to fibre diameter.	Fine fibres and shaft portions of guard fibres are oval to rectangular with a narrow cortex surrounding one or more medullae. The shield portion is irregular, often flattened and has a number of medullae.	RABBIT and HARE FUR Figs 26–28
Coarse, regular in diameter with smooth profile.	Oval to circular. Generally concentric medulla of the wide lattice type. Pigment—frequently concentrated near medulla.	HORSE HAIR (from mane) Figs 30, 31
	Oval to circular. Pigment—sparse to very dense. Medulla—stellate, oval or circular.	HORSE HAIR (from tail) Figs 29, 31
Regular in diameter with fairly prominent scale margins. Pigment—sparse to dense. Medulla—many fibres non-medullated and even in the coarser fibres the medulla is relatively narrow.	Oval, medulla when present is relatively narrow, some flattened and possibly unpigmented with concentric medulla.	COW HAIR (from body) Figs 33, 34
Coarse, regular in diameter with smooth profile. Pigment—very sparse. Medulla—when present is often narrow and interrupted.	Circular to oval. Fibres up to 200 μm diameter. Cortical cell boundaries can often be seen.	COW HAIR (from tail) Figs 32, 33
Fine fibres—oval or circular with thin cuticular layer. Pigment granules may be arranged bi-laterally. Coarse fibres—oval to circular with fairly thick cuticular layer. Medulla when present is narrow. Pigment varies from none to very dense and is usually evenly distributed.	Fine fibres—regular diameter with slightly prominent scales. Non-medullated with very sparse to dense pigmentation. Coarse fibres—regular diameter with smooth profile and shallow scales. Medulla is narrow and sometimes discontinuous.	YAK HAIR Figs 35, 36

TABLE VII—*continued*

Longitudinal View	Cross-sectional View	The Specimen is
Fine fibres or filaments, cemented in pairs by silk gum. The gum layer is not always continuous.	Triangular, with rounded corners, in pairs.	**RAW BOMBYX SILK** Fig 38
Fine fibres or filaments variable in diameter, single, smooth, nearly structureless, sometimes flattened. Occasionally very fine fibres are seen, formed by superficial splitting of the original fibre.	Separated fibres, triangular, rounded corners.	**DEGUMMED BOMBYX SILK** Figs 37, 39–42
Flat irregular ribbons, usually separate, sometimes twisted and with longitudinal striations.	Very elongated triangles, usually separate.	**TUSSAH SILK** Figs 43, 44
Fibres uniform in diameter along their length, structureless or nearly so. Possibly slight variation in diameter from fibre to fibre.	Circular or nearly so; sometimes with flats or indentations. Any variation in size is clearly seen in cross-section.	**REGENERATED PROTEIN FIBRES**

Note. All animal fibres have cuticular scales but if the fibre is dyed to dark colours, is densely pigmented, or has a wide medulla, the scale pattern will not be seen. The presence of scales will be detected on the profile of the fibre.

Protein Fibres: Identification by Solubility

Bombyx silk, tussah silk, Merinova, wool and hair.

TABLE VII (*a*)

Treat the fibres in a test-tube with the following reagents, using a fresh portion of the fibre for each test. Continue with the reagents in succession, and in the order given, until one is found in which the fibre dissolves.

First Reagent of the Series in which the Specimen Dissolves	The Specimen is
Calcium chloride in formic acid (see Section 5.17.3)	BOMBYX SILK
Concentrated sulphuric acid	TUSSAH SILK
Trypsin (see Section 5.17.15)	MERINOVA
Insoluble in all these reagents	WOOL and HAIR

Confirmatory Tests for Protein Fibres

Wool and hair are stained dark brown in alkaline lead acetate solution (see Section 5.17.1). Other fibres are unstained.

Tussah silk, degummed, is only lightly stained in Millon's reagent (see Section 5.17.10), wool and hair, degummed *Bombyx* silk, and Merinova are deeply stained.

4.2.3.2   Cellulose Fibres

Bast, leaf, and fruit fibres, comprising flax, ramie, abaca (manila), *Phormium tenax*, sisal, hemp, sunn fibre, jute, *hibiscus, urena* fibre, coir; kapok, akund; cotton, mercerized cotton, chemically modified cotton; regenerated cellulose, continuous-filament and staple, with or without chemical modification.

TABLE VIII

Examine the fibres microscopically in longitudinal view mounted (*a*) in liquid paraffin, (*b*) in water; and in cross-section.

TABLE VIII

Longitudinal View	Cross-sectional View	The Specimen is
Fibres in bundles of ultimates or mixtures of bundles of various sizes and single ultimate fibres. Often cross-markings and dislocations in ultimates.	Bundles often composed of polygons.	**BAST, LEAF, and FRUIT FIBRES** Figs 55–81. Proceed as in Table VIII (a)
Smooth and cylindrical with rounded base; no convolutions or other structure except at the ends. When mounted in water show large, elongated air-bubbles.	Circular or slightly elliptical, thin wall, and large lumen.	**KAPOK and AKUND** Figs 51–54
Fibres separate. Flattened and ribbon-like with frequent convolutions that sometimes change direction. Some fibres (immature fibres) have thin walls and few convolutions. Tips of fibres rounded, bases irregular and torn. (Mercerized cotton is smoother than unmercerized. Convolutions and lumen are less prominent and may even be absent from some fibres[1].)	Flat, elongated, or bean-shaped, with the lumen as a line parallel to the longer direction, some very thin-walled sections of immature or dead fibres. (Well mercerized cotton has circular or elliptical shaped cross-sections, with the lumen appearing as a small central spot[1].)	**COTTON** Figs 47–50. Proceed as in Table VIII (b)
Appearance consistent with origin by extrusion, i.e., uniform diameter, lack of structural interruptions along the length, striations (if present) are parallel and continuous, pigment particles (if present) are evenly distributed throughout the mass.	Relatively uniform shape within one particular type. Possible shapes include round, round with slight flattening, bean, serrated (common type), folded, or hollow.	**REGENERATED CELLULOSE** Figs 89–102. Proceed as in Table VIII (d)
Regular, uniform, structure which may show striations. Longitudinal view is dominated by discrete (oily) globules which are evenly distributed within the fibres and may protrude at the fibre surfaces.	As for regenerated cellulose. Possibly with some evidence of inclusions in the fibre.	**FLAME-RETARDANT VISCOSE** Figs 99, 100. Proceed as in Tables VIII (d) and VIII (h)

[1]The extent of mercerization of individual samples of cotton varies widely. Most samples of mercerized cotton will have fibres of both swollen and unswollen types.
Mercerized cotton and Organdie. Congo Red Test (Section 5.17.5). Mercerized cotton fibres are stained faint pink and do not adhere. Organdies are stained bright red and the fibres adhere to some extent.

Examination of Bast, Leaf, and Fruit Fibres

Flax, ramie, abaca (manila), *Phormium tenax*, sisal, hemp, sunn fibre, jute, *Hibiscus* fibre, *Urena* fibre, coir.

TABLE VIII (*a*)

The fibres are illustrated in longitudinal view and in cross-section in Figs 55–81 where attention is drawn to salient morphological features. This table should be used in conjunction with the illustrations, or with authentic specimens of the fibres in textile form examined under the microscope.

Apply Twist Test (see Section 5.12)	Examine Plate Cross-section at 50 to 100× (see Section 4.1.5)	Apply Billinghame's Test (see Section 5.10)	Determine Mean Length of Ultimate Fibres (see Section 5.9)	Examine Epidermal Tissue (see Section 5.18)	Examine Ash (see Section 5.13)	The Specimen is
Clockwise	—	—	< 50 mm	—	—	FLAX Figs 55–57
			> 50 mm	—	Cluster crystals Fig. 78 (*a*)	RAMIE Figs 66, 67
Counter-clockwise	Frequently 100 ultimates in each fibre bundle. Circular, elliptical, or crescent-shaped bundles	Orange	> 4 mm	—	Stegmata are present Fig. 81(*a*)	ABACA (MANILA) Figs 71, 72, 81 (*a*)
		Yellow	> 4 mm	—	—	PHORMIUM TENAX Figs 73, 74
			< 4 mm	—	Rod-like crystals Fig. 80 (*b*)	SISAL Figs 68–70, 80 (*b*)

TABLE VIII (a)—(continued)

Apply Twist Test (see Section 5.12)	Examine Plate Cross-section at 50 to 100× (see Section 4.1.5)	Apply Billinghame's Test (see Section 5.10)	Determine Mean Length of Ultimate Fibres (see Section 5.9)	Examine Epidermal Tissue (see Section 5.18)	Examine Ash (see Section 5.13)	The Specimen is
		—	> 8 mm	Pigmented cells, short hairs	Cluster crystals Fig. 80 (a)	**HEMP** Figs 61, 62, 65 (a), 80 (a)
	Usually less than 30 ultimates in each bundle. Bundle shapes irregular		> 3.7 mm	Clear cells, long hairs	—	**SUNN FIBRE** Figs 63, 64, 65 (b)
Counter-clockwise		—	< 3.7 mm		Solitary crystals Fig. 79 (a)	**JUTE** Figs 58–60, 79 (a)
				—	Cluster crystals Fig. 79 (b)	**HIBISCUS FIBRE or URENA FIBRE** Fig. 79 (b)
Indecisive	—	—	< 1.5 mm	—	Stegmata are present Fig. 81 (b)	**COIR** Figs 75–78, 81 (b)

Confirmatory Tests
Jute: See Section 5.11.
Urena fibre: Acetyl content, 98–124 milliequivalents per 100 g fibre.

Examination of Cotton and Chemically Modified Cotton

TABLE VIII (*b*)

Add a drop of cuprammonium hydroxide to the fibres on a slide, and examine under the microscope.

Behaviour of Specimen	The Specimen is
Fibres dissolve readily and completely	SCOURED COTTON or MERCERIZED COTTON
Fibres swell irregularly, forming beads and finally dissolve leaving a residue of particles that float away in the solvent	RAW COTTON
Fibres remain undissolved. (Some samples may swell and stain blue, especially at fibre tips)	CHEMICALLY MODIFIED COTTON

Differentiation of Chemically Modified Cotton

Chemically modified cotton includes acetylated cotton, resin-treated cotton, and cross-linked cotton.

Acetylated cotton may be distinguished from resin-treated cotton and cross-linked cotton as follows.

(*a*) Put the sample in a test-tube containing cuprammonium hydroxide. Acetylated cotton dissolves in 2–3 hours, but resin-treated cotton and cross-linked cotton do not.

(*b*) Boil the sample in 0·5N alcoholic potash for 30 min. With acetylated cotton the residue, after washing with water and removing the surplus, dissolves in cuprammonium hydroxide, but with resin-treated cotton and cross-linked cotton, it does not.

If the sample is resin-treated cotton or cross-linked cotton, proceed as in Section 5.3.

Examination of Regenerated Viscose and Cuprammonium Fibres

TABLE VIII (*c*)

Mount the fibres in liquid paraffin (*n* 1·47) and examine under the microscope in longitudinal view at a magnification of 200–400×. Cut sections by the plate method and observe their shape.

Refer also to the confirmatory data in Table VIII (*g*).

TABLE VIII (c)

Microscopical Examination		Shirlastain A (See Section 5.17.7)	Stained Cross-sections[1] (See Section 5.6)		Form	Fibre Type
Longitudinal	Cross-section					
Rod-like	Off-round Fig. 102	Blue	—	—	Filament	CUPRAMMONIUM
	Round to Bean	Pink	Thin, even skin Fig. 96		Staple	HIGH-WET-MODULUS VISCOSE e.g. Vincel 64
			All skin Fig. 92		Filament	HIGH-TENACITY VISCOSE e.g., Tenasco Super 2A
Striated	Serrated	Pink	Thin, even skin Fig. 94		Staple	IMPROVED-TENACITY VISCOSE e.g., Fibro HS
			Thin, even skin Figs 91, 93		Filament and staple	REGULAR VISCOSE e.g., Textile, Fibro
			Uneven skin Figs 97, 98		Staple	CRIMPED VISCOSE[2] e.g., Sarille (fine), Evlan (coarse)
Striated, globules visible	Serrated	Pink	Even or uneven skin Fig. 99		Filament and staple	FLAME-RETARDANT VISCOSE e.g., Darelle

[1]Examination of 'skin' requires thin cross-sections. See Section 4.1.5.3.
[2]Crimped viscose curls rapidly when placed in a 3% solution of caustic soda. Regular viscose tends to straighten out.

Identification of Chemically-treated Viscose Fibres

TABLE VIII (*d*)

Treat the fibres with cuprammonium hydroxide on a microscope slide and examine under the microscope.

Observation	The Specimen is
Fibres dissolve	UNTREATED VISCOSE
Fibres do not dissolve	CHEMICALLY-TREATED VISCOSE e.g., resin-finished or flame-retardant-finished

Examination of Cellulosic Fibres which do not Burn Readily

TABLE VIII (*e*)

If necessary remove any finish with dilute hydrochloric acid (See Appendix A) and strip dyestuff in hydrosulphite (See Section 5.1). Re-dye in Celliton Blue FFR (See Section 5.17.4) and rinse thoroughly. Examine microscopically.

Appearance	Specimen is
Blue globules visible throughout fibre	FLAME-RETARDANT VISCOSE
Unstained or very lightly stained	VISCOSE WITH FLAME-RETARDANT FINISH

Identification of Deep-dyeing Viscose

TABLE VIII (*f*)

If necessary remove any finish with dilute hydrochloric acid (0·02N) and strip dyestuff in hydrosulphite (See Section 5.1). Re-dye in Lissamine Red 2G/Solophenyl Blue Green BL (See Section 5.17.8) and rinse thoroughly. Examine microscopically.

Appearance	Specimen is
Fibres dyed pink	DEEP-DYEING VISCOSE
Fibres dyed blue–green	NORMAL-DYEING VISCOSE

Note. Unless finish is removed adequately before re-dyeing, flame-retardant-finished viscose may dye red–purple under these conditions. This will be clear from Tables VIII (*d*) and (*e*). If necessary, repeat extraction using 0·1N hydrochloric acid at 60 °C for 30 min.

Table of Confirmatory Data (Regenerated Cellulose Fibres)

TABLE VIII (g)

Type of Regenerated Cellulose	Examples of Trade Names	Linear Density Range (dtex)	Microscopical appearance		Birefringence	Fluidity[1]	Tensile Properties				Water[2] Imbibition (%)
			Longitudinal	Cross-sectional			Tenacity (mN/tex)		Extension (%)		
							Air Dry	Wet	Air Dry	Wet	
Continuous filament											
(i) Regular (textile)	Textile	66–2220	Striated	Serrated	0·022	10–11	190	90	20	27	85–95
(ii) High tenacity (tyre yarn)	Tenasco Super	1220–2440	Rod-like	Smooth, bean to circular	0·035–0·040	4–5	420	320	12	22	70–75
(iii) Flame retardant (textile)	Darelle	330	Striated with globules	Serrated	0·020	10–11	130	60	19	25	75–85
(iv) Cuprammonium	Cupresa	28–330	Rod-like	Smooth, almost circular	0·034	4	160	80	13	19	90–110
Staple fibre											
(i) Regular	Fibro	1·7–5·0	Striated	Serrated	0·024	8–11	210	115	19	24	95–105
(ii) Improved tenacity	—	1·3–1·7	Striated	Smoothly indented	0·030	8–10	240	140	17	22	90–95
(iii) High tenacity	Durafil	1·7	Rod-like	Smooth, almost circular	0·035	4–5	330	240	22	30	65–70
Crimped staple											
(i) Fine	Sarille	2·4–5·0	Striated	Serrated	0·027	8–10	200	110	24	32	85–95
(ii) Coarse (carpets)	Evlan	9·0–17·0	Striated	Serrated	0·020	8–10	175	85	26	38	85–95
Flame-retardant staple											
(i) Regular	Darelle	5·0	Striated with globules	Serrated	0·023	8–11	150	85	22	29	75–85
(ii) Crimped	Darelle	9·0–17·0	Striated with globules	Serrated	0·020	8–11	130	65	20	26	75–85
High-wet-modulus staple	Vincel 64	1·7–5·0	Rod-like	Smooth, bean to circular	0·038[3]	4–5	380	240	15	18	70–75

[1]Fluidities measured according to BS 2610 using a 2% solution in cuprammonium hydroxide.
[2]Water imbibition is the amount of water retained after a wet sample has been extracted at 1000 g for 5 min expressed as a percentage of the dry weight of the sample.
[3]High-wet-modulus fibres are available in the birefringence range 0·30–0·47.

4.2.3.3 *Calcium Alginate Fibres*

TABLE IX

Examine the fibres microscopically in longitudinal view and in cross-section.

Longitudinal View	Cross-sectional View
Uniform in diameter along length, not necessarily uniform from fibre to fibre. Striations parallel to length	Irregular and elongated with very jagged edges

Confirmatory Tests for Calcium Alginate Fibres

1. Place a small tuft in carbon tetrachloride (density 1·6 g/cm³).
   —Calcium alginate (density 1·7 g/cm³) sinks.
2. Boil in 2% caustic soda.
   —Calcium alginate becomes a bright yellow.
3. Warm in ten times its weight of 3% sodium carbonate; decant the liquid, concentrate to a small volume, and acidify with sulphuric acid.
   —Calcium alginate dissolves and alginic acid is precipitated as a colourless gel on acidification.

4.2.4 Non-combustible Fibres (Table X)

Glass, asbestos

TABLE X

Appearance, using Microscope where needed, Longitudinal View, Dry Mount		Flame Test	Indication
Length and Diameter	**Character**		
Continuous-filament, uniform diameter	Rod-like	Bright orange	A-GLASS FIBRE or SODA-LIME BOROSILICATE FIBRE Distinguish by analysis (see p. 43)
		Pale yellow to colourless	E-GLASS FIBRE or TEMPERATURE-RESISTANT FIBRE Distinguish by analysis
		Vivid blue flame	LEAD GLASS
Staple, irregular diameter	Curved 'shot'[1] present	Orange–yellow	INSULATING WOOL
Staple, more regular diameter	Rod-like	Orange–yellow	A-GLASS FIBRE or SODA-LIME BOROSILICATE FIBRE Distinguish by analysis

[1]Globule of undrawn glass.

Dressings, Sizes, and Finishes on Glass Fibre

The presence of a dressing or size may be detected by the hot-plate test of Table I. Smoke and perhaps discoloration are observed when a dressing or size is present. The components may be partially separated by solvent extraction, and identified by infrared spectroscopy and chromatography. If no fuming occurs the specimen may have been desized or finished. Silane finishes can be distinguished by colour reactions based on comparison with known specimens. Chrome finish is susceptible to general chemical identification and estimation. A uniform brown colour indicates one of the 'caramelized' finishes.

Differentiation of Glass Fibres

To confirm the distinction between A-glass and E-glass, a refractive index test may be used. Immerse the fibres in methyl salicylate on the microscope stage using the best illumination and condenser for Becke-line observation (see Section 5.4). E-glass has a higher refractive index than methyl salicylate, and high-alkali glass has a lower refractive index than methyl salicylate.

Differentiation of Asbestos Fibres

Chrysotile

Several methods may be employed to identify chrysotile, and preferably more than one should be utilized with any specific sample in order to obtain absolute proof of identity. This is particularly important either when chrysotile fibres are present in admixture with other materials, where prior separation may be necessary, or if the quantity of specimen available is small. The principal methods for identifying chrysotile are summarized below.

X-ray Diffraction

Chrysotile gives a unique diffraction pattern enabling a positive identification to be made easily. This is the most reliable method for both the identification and the quantitative estimation of chrysotile in a mixture of other materials, and particularly when the quantity of available sample is small.

Principal d-spacings are given in the ASTM index and are reproduced in the following table.

Principal d-Spacings of Clino-chrysotile (ASTM 10–381)

d(Å)	I/Io	h kl
7·36	100	002
3·66	100	004

Electron Microscopy

Identification can be based on the observation of the unique morphology of chrysotile fibrils and by the electron diffraction pattern (Fig. 88). Suspected chrysotile specimens can be broken down into their ultimate fibrillar form for electron microscopic examination by suspension in a surfactant solution which is then subject to ultrasonic treatment. Submicrogram quantities of chrysotile can be estimated by a method developed by Rickards[18].

---

[18]A. L. Rickards. *Anal. Chem.*, **45**, 809.

Infrared Spectroscopy
Chrysotile has a characteristic infrared spectrum in which the vibrations from structural hydroxyl groups at 2·71 and 2·74 μm, and from silicon oxygen bonds near 9·4, 9·8, and 10·4 μm, can clearly be seen in the 2·5–15 μm region (Diag. 4.5). Further confirmation can be gained from characteristic bands present between 15 and 50 μm if a suitable instrument is available. The presence of large quantities of other silicates can make positive identification difficult.

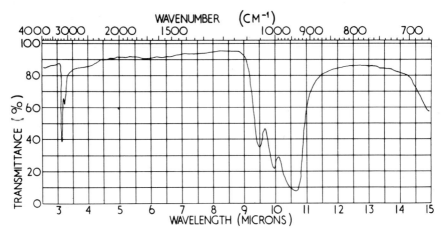

4.5 Infrared spectrum of chrysotile, 2·5–15 μm.

Differential Thermal Analysis
On heating from ambient to 1000 °C, the dehydroxylation of chrysotile can be observed by a large endothermic peak which commences at approximately 550 °C and ends at near 800 °C. The most characteristic feature is a peak at approximately 820 °C which is due to the exothermic formation of forsterite (Diag. 4.6). The exact shape and location of the peaks is affected by the degree of subdivision of the fibres.

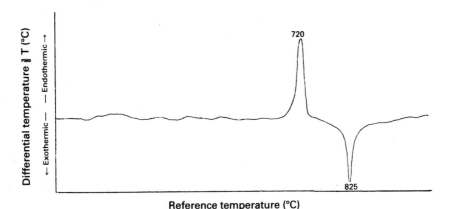

Diag. 4.6 DTA curve of 100 mg chrysotile.

Thermogravimetric Analysis
Programmed heating to 1000 °C shows a weight loss of approximately
13% in the 550–800 °C region due to dehydroxylation (Diag. 4.7). The
exact shape and location of the peak is affected by the degree of subdivision of the fibres.

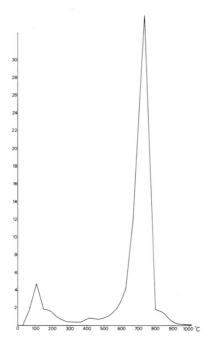

Diag. 4.7 First derivative from thermogram of 609 mg chrysotile.

Light Microscopy
Until recently this has been the technique most frequently used; identification is based on refractive index (*n* 1·50–1·56) and measurement of
shape or form[19]

Determination of Silicon and Magnesium
If other materials containing these two elements are absent, a quantitative
determination of silicon and magnesium may give useful confirmatory evidence for the presence of chrysotile. Typical values are MgO 38–41%,
SiO_2 39–42%.

Neutron Activation Analysis
This technique has been used to detect chrysotile through the presence of
trace amounts of scandium which appears to have a close association with
chrysotile.

Amphiboles
These fibres are most readily detected by x-ray diffraction, electron
diffraction (more complex) (Figs. 86, 87), or infrared spectroscopy.
    Principal d-spacings for crocidolite and amosite are given below and
infrared spectra are shown in Diags 4.8 and 4.9.

---

[19]S. Holmes. *Ann. N.Y. Acad. Sci.*, 1965, **132**, 288.

Light microscopy, may also be useful, particularly in the case of crocido-lite, because of its blue colour, but the refractive indices of amosite (*n* 1·64–1·69) and crocidolite (*n* 1·68–1·71) are too close to enable distinc-tions between these two to be made by this method.

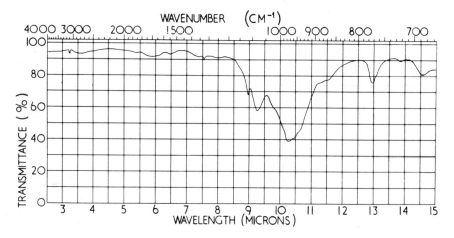

Diag. 4.8 Infrared spectrum of crocidolite, 2·5–15 μm.

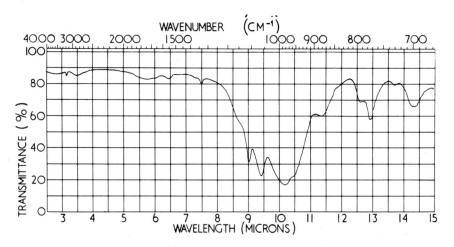

Diag. 4.9 Infrared spectrum of amosite, 2·5–15 μm.

Principal d-Spacings of Crocidolite (Riebeckite ASTM 19–1061)

d(Å)	I/Io	h kl
9·02	4	020
8·40	100	110
4·51	16	040,021
*3·12	55	310

Principal d-Spacings of Amosite (Grunerite ASTM 17–725)

d(Å)	I/I	h kl
9·21	50	020
8·33	100	110
4·68	30	200
4·58	35	040
4·16	40	220
3·28	50	240
*3·07	80	060,310
2·766	90	151

*Main diagnostic line.

4.2.5 Metal–foil Yarns (Table XI)

Initial Burning Test

Burn the specimen in a small flame and observe the following:

(a) Yarn burns leaving a metal foil—foil type is indicated—proceed as in Table XI.

(b) Yarn burns leaving no metal foil—metallized type is indicated—proceed as in the Notes below.

Notes

1. Treatment with warm acetone dissolves the adhesives used in the construction of polyester foil laminates and metallized polyester laminates and separates the yarn into two or three plys according to type.

2. Differentiation between aluminium foil and metallized polyester cores can be achieved by treatment with 30% hydrofluoric acid[20]—foil dissolves completely, whereas in the case of metallized film a clear polyester film layer remains.

3. If no separation into layers occurs the yarn is a non-laminate. Colour and lacquer, or both, are usually removed from non-laminates by the acetone test to expose a metallized surface on one or two sides.

4. Alternatively, microscopical examination of cross-sections of yarn can be used to differentiate between laminate and non-laminate yarns.

---

[20]*Warning* Hydrofluoric acid can cause serious skin burns. It readily attacks metals, glass, and glazed surfaces and should be used from polythene, gutta percha, or Bakelite containers.

TABLE XI

Burning Test	Hand-stretching Test	Boiling Water (1 min)	Acetone (5 min)	Cold Shirlastain A	75% Sulphuric Acid	Conc. Sulphuric Acid	Boiling Xylene	Film Type
Film does not melt; burns readily with sizzle; burnt-paper smell	Stretch < 20%	Yarn delaminated	Yarn sometimes delaminates	Pink	Film dissolves	Film dissolves	—	CELLULOSE e.g., Cellophane
Film burns readily with grey-blue smoke; smells of rancid butter	Stretch < 50%	Unaffected	Film dissolves	Lime green	Film dissolves	Film dissolves	—	CELLULOSE ACETATE BUTYRATE
Film burns with black sooty smoke; may be self-extinguishing	Stretch > 80%	Unaffected	Yarn usually delaminates	Unstained	Unaffected	Unaffected	—	POLYESTER
	Stretch 50–80%	Unaffected	Yarn usually delaminates	Unstained	Unaffected	Unaffected	Film dissolves	POLY-PROPYLENE

4.2.6   Preliminary Sorting Tests by Solubility (Table XII)

This method is not intended to replace the preliminary sorting tests by heating (see Section 4.2.1), which form the introduction to the main scheme of analysis, but some readers may find it a useful alternative.

TABLE XII

Treat the specimen in a test tube with the first reagent in the left-hand column and proceed as indicated (see next page).

4.3   Fibre Blends

Many textile materials contain two or more fibre types. Additional types of fibre may be introduced as substitutes on economic grounds, to create decorative effects, to produce materials with improved properties or special effects, to take advantage of differences in dyeing properties of the fibres, or to facilitate the production of fabrics of a special type.

Mixtures containing more than one type of fibre may be produced by combining in various ways two or more yarns, each composed wholly of a different type of fibre, which may be either in staple or in continuous-filament form. A mixture of this type may often be recognized by differences in the appearance of the component yarns, but when uncertainty exists, a microscopical examination of the component yarns should be made. The yarns of different type may then be separated by hand and each subjected to the test appropriate for the analysis of materials of homogeneous composition (see Tables I–XI).

On the other hand, many textile materials consist wholly or partly of yarns containing two or more fibre types blended together in staple form prior to spinning, and methods for the identification of the component fibres are of real importance. Yarns of the Tricelon type, i.e., a mixture of Tricel and nylon, are composed of separate filament yarns subsequently combined with or without twist and the individual components are not very obvious. Hand separation of such yarns is likely to be difficult and solubility tests using the microscope can be used with advantage.

It is first necessary to establish that more than one type of fibre is present and this is most satisfactorily done by microscopical examination, both in longitudinal view and in cross-section, making observations of such features as scales, convolutions, markings, the presence of delustrants and differences in colour, size, and shape (see Section 4.1). When the specimen is undyed or dyed to a pale shade, preliminary staining with a suitable selective stain, e.g., Shirlastain A, is often helpful.

The methods given in the main scheme of analysis (Tables I–XI) are intended for application to homogeneous specimens only and are unsuitable for use with blends, and an alternative method of analysis must be used. Probably the most effective method is to make use of the solubilities of fibres in different reagents, and a suitable scheme is given in Table XIII. In this scheme, grouping reagents are first applied in the order given to the same specimen to classify each component in turn into a solubility group. Other reagents, listed along the top of the Table, are then used in the order given for the appropriate group, from left to right, in order to identify the fibre finally. It will be noted that bast and leaf fibres have not been included, although bast fibres are sometimes encountered in blends.

TA

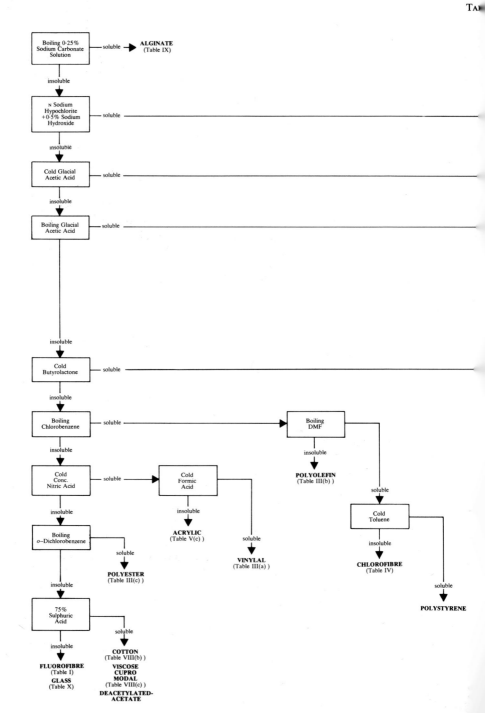

Boiling 0·25%
Sodium Carbonate
Solution — soluble → **ALGINATE**
(Table IX)

insoluble

N Sodium
Hypochlorite
+0·5% Sodium
Hydroxide — soluble

insoluble

Cold Glacial
Acetic Acid — soluble

insoluble

Boiling Glacial
Acetic Acid — soluble

insoluble

Cold
Butyrolactone — soluble

insoluble

Boiling
Chlorobenzene — soluble → Boiling
DMF

insoluble   insoluble

Cold
Conc.
Nitric Acid — soluble → Cold
Formic
Acid

POLYOLEFIN
(Table III(b))

soluble

insoluble   insoluble   Cold
Toluene

Boiling   **ACRYLIC**   soluble
o–Dichlorobenzene   (Table V(c))

soluble   **VINYLAL**   insoluble
(Table III(a))

POLYESTER   **CHLOROFIBRE**
(Table III(c))   (Table IV)

insoluble   soluble

75%
Sulphuric   **POLYSTYRENE**
Acid

soluble

insoluble

COTTON
(Table VIII(b))

FLUOROFIBRE   **VISCOSE**
(Table I)   **CUPRO**
MODAL
GLASS   (Table VIII(c))
(Table X)
DEACETYLATED-
ACETATE

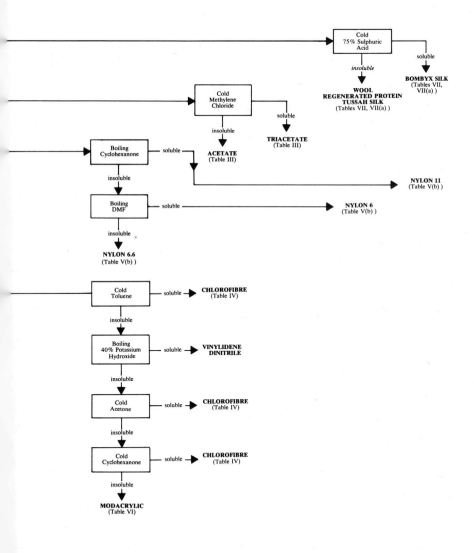

	Grouping Reagents	Fibre Type	Solvents for Group 2					S...
			70% v/v Acetone	Glacial Acetic Acid	4·4N Hydrochloric Acid	35% w/w Sulphuric Acid	5N Hydrochloric Acid 65°C	Tr... T... (... Se 5....
1	Fibre soluble in boiling 0·25% sodium carbonate	CALCIUM ALGINATE						
2	Fibres soluble within 15 min in calcium chloride–90% formic acid 1:10 (see Section 5.17.3)	DIACETATE	SOL 1 min	sol				
		TRIACETATE		SOL 4 min				
		NYLON 6			SOL 2 min	sol	sol	
		NYLON 6.6				SOL 2 min	sol	
		VINYLAL					SOL 5 min	
		BOMBYX SILK						
3	Fibres soluble within 5 min in N sodium hypochlorite + 0·5 sodium hydroxide (see Section 5.17.13	REGENERATED PROTEIN						1
		WOOL						
		TUSSAH SILK						
4	Fibres soluble within 8 min in 1,4–butyrolactone	COPOLYMER OF VINYL CHLORIDE–VINYL ACETATE						
		CHLORINATED POLY(VINYL CHLORIDE)						
		MODACRYLIC						
5	Fibres soluble within 10 min in conc. sulphuric acid	NYLON 11						
		ACRYLIC						
		REGENERATED CELLULOSE						
		BLEACHED COTTON						
		ACETYLATED COTTON						
		POLYESTER						
6	Fibres insoluble in all the above reagents	POLY(VINYL CHLORIDE)						
		COPOLYMER OF VINYLIDENE CHLORIDE–VINYL CHLORIDE						
		POLYOLEFIN						
		SYNDIOTACTIC POLY(VINYL CHLORIDE)						
		FLUOROCARBON						

Solvents to be used at room temperature unless otherwise stated.

III

Chloroform	Commercial Xylene	meta-Cresol	Conc. Nitric Acid	60% w/w Sulphuric Acid 60°C	Cuprammonium Hydroxide	75% w/w Sulphuric Acid	Tetra hydrofuran	Dioxan 80°C	Commercial Xylene Boil	Form-dimethylamide Boil	
						sol					**1**
		sol	sol	sol		sol	sol	sol		sol	
sol	sol	sol	sol	sol		sol		sol		sol	
		sol	sol	sol		sol				sol	**2**
		sol	sol	sol		sol					
			sol	sol		sol					
			sol	sol	sol	sol					
				sol	sol						**3**
SOL ½ min	sol							sol	sol	sol	
	SOL 5 min							sol	sol	sol	**4**
										sol	
		SOL 3 min		gels		gels				sol	
			SOL 3 min			sol				sol	
				SOL 7 min		sol					**5**
					SOL 15 min	sol					
						SOL 20 min					
										sol	
							SOL 1 min	sol	sol	sol	
								SOL 1 min	sol	sol	**6**
									SOL 2 min	gels	
										SOL ½ min	

Blank spaces denote fibres insoluble within 30 minutes.

The information given in Table XIII can be used in the two ways outlined below, the choice depending on the size of the specimen and whether or not the reagent requires heating. The solubility times quoted in the Table are an indication of the rate at which the fibre dissolves and it should be clearly understood that these data relate to the test-tube method only; solubility times on a microscope slide are not necessarily the same. Blank spaces denote 'insoluble within 30 minutes'.

Although this scheme is primarily intended for application to fibre blends, it may be used as an alternative to the main scheme for the identification of homogeneous specimens (Tables I–XI) when facilities for carrying out the tests required for the latter scheme are not available. For this reason, some fibres which are not usually found in blends are included.

When possible, the observations noted in this scheme should be supplemented by intelligent use of the confirmatory tests given in Tables III–XI. In order to apply confirmatory tests, the components should first be separated, e.g., by making use of either a density or a solubility difference.

Recommended Procedure

Group separation
Treat the specimen in a test-tube for the time stated with the second reagent, i.e., calcium chloride–formic acid, given in the left-hand column of Table XIII. The first reagent may usually be omitted, because alginate fibres are not normally found in blends. They may however, occur in waste material and, if their presence is suspected group reagent 1 should first be used. Collect the undissolved fibre residue (if any) with a sintered-glass filter (porosity 1) and wash it with the reagent, followed by water, and dry it with filter paper. Determine if one (or more) component(s) has dissolved by examining the residue under the microscope or by examining the solution (see *Note* (i), p. 187). Then proceed as indicated in the following table.

Behaviour of Specimen	Further Procedure
The specimen disolves completely in the group reagent	Using a new specimen, proceed with *Component Identification* as given below
One (or more) of the components dissolves, but an insoluble fibrous residue remains	(i) Using a new specimen, proceed with *Component Identification* of the soluble component(s) as given below (ii) Using *half* the residue, proceed with *Group Separation*, as above, using the next group reagent (iii) Retain the remainder of the residue and use it for the next component identification N.B, do *not* use a fresh specimen
The specimen is unaffected, i.e., no component is dissolved	Using the residue, proceed with *Group Separation*, as above using the next group reagent

Component Identification
If any component dissolves in the group 2 reagent, it is classified as a member of group 2. To identify this component, treat a new specimen of the blend with the first solvent listed under group 2 in the second row of the table, i.e., 70% acetone. Observe if a component dissolves in this solvent. Treat the residue with the next solvent in order and proceed in this way

until a solvent is found in which the group 2 component(s) dissolves. Each component is then identified as the fibre named in the corresponding horizontal row.

Further Procedure
Repeat the above procedure, group by group, on the residues, in the order given, until each component of the blend has been identified.

If fibres remain undissolved after treatment with concentrated sulphuric acid (group 5 reagent), this residue should be subjected to the group 6 identification reagents.

Notes (i) When there is any doubt that a component has in fact dissolved, e.g., when two or more fibres having similar microscopical appearance are present, confirmation may often be obtained in the following ways.

(*a*) If the reagent is an organic solvent, evaporate the filtrate to dryness and note if any residue, usually in the form of a film, is present.

(*b*) If the reagent is an acid, dilute the filtrate; if any cloudiness develops, the filtrate contains a dissolved fibre.

(ii) When heating is specified for a treatment with a solvent, place the test-tube containing the fibre and reagent in a glass beaker water-bath at the required temperature. Where boiling is specified apply a direct flame to the test-tube with the usual precautions. Agitate the contents of the test-tube by gentle shaking.

(iii) If at any stage in this scheme of analysis a residue is found containing one type of fibre only, the methods applicable to homogeneous specimens (Tables I–XI) can be applied.

(iv) A blank space in Table XIII indicates that the fibre is insoluble in the corresponding solvent within 30 minutes. Where the dissolution of a fibre in a particular solvent serves to identify the fibre unequivocally, the abbreviation 'SOL' is printed in heavy type. The abbreviation 'sol' in normal type means that the fibre is soluble in the corresponding solvent and the fact affords confirmatory, but not decisive, evidence of identification.

Alternative Procedure for Observing Behaviour in Reagents

This procedure is unsuitable when heating is necessary.

When the amount of material available for examination is very small, the effect of some of the reagents used in the above scheme may be observed under the microscope. First mount the fibres dry on a slide, bring them into focus, and observe them while the slide is irrigated with the reagent. Care must be taken to avoid incorrect observation, firstly because the fibres usually dissolve more slowly under a cover glass and, secondly, because some undyed fibres become invisible (optically dissolved) when mounted in certain reagents, because the refractive indices of fibre and reagent are similar. This phenomenon must not be mistaken for true solution and if this is suspected the fibres should be stained or dyed prior to examination.

4.4   Instrumental Methods of Analysis

4.4.1   Infrared Spectroscopy

The identification of polymers in general and synthetic fibres in particular can be achieved readily by this technique, but, for separation into class and identification of the commoner fibres, this method is an unnecessary refinement since microscopical, chemical, and staining tests will usually suffice. Furthermore, if two or more synthetic fibres are derived from the same basic monomer, whose properties have been modified by the addition of the same copolymer in different amounts, and if the percentage difference is small, it may not be possible to distinguish the fibres by qualitative infrared examination. Where the copolymer is different, however, then the infrared spectrum obtained will be specific for that particular fibre. One great advantage of infrared examination is that the spectrum obtained is determined mainly by the chemical constitution of the fibre and is, in general, less dependent on physical structure, variations in which can affect the results obtained from staining, solubility, and other physical tests used for fibre identification. Where only a few milligrams of sample are available, infrared spectroscopy is probably the most valuable single test. The method is particularly useful with the newer synthetic fibres such as polyolefin and acrylonitrile fibres, especially the latter, where the constitution and proportion of the copolymer used are frequently modified. Care should be taken in interpreting spectra of acrylic fibres as a peak (at 5·98 μm) due to residual form–dimethylamide from manufacture may be noticed.

When infrared radiation is passed through a substance, certain frequencies are absorbed and others are transmitted; this is, of course, similar to the absorption of visible light which causes materials to show colour. Infrared spectroscopy, therefore, consists of determining the frequencies at which absorption occurs and preparing a plot of percentage radiation absorbed against frequency. In practice, this is carried out automatically by the spectrometer.

Although it is not intended to describe the various instruments available, or the origins of infrared spectra, it is worth noting that the majority of commercial double-beam spectrophotometers scan the spectrum from 2 to 15 μm and record the difference between the energies in the two beams on a suitable chart[1,2]. Because of the number and complexity of the absorption bands, the infrared spectrum of a given molecule is characteristic of that compound and may be used for identification. In comparative studies of two substances, therefore, identical infrared spectra denote identical substances.

4.4.1.1 Procedure

The technique of operating the instrument is not essentially any different for fibre identification than for the examination of other organic compounds[3].

The spectra of relatively simple organic molecules are usually determined with the compound itself or in a medium transparent to infrared radiation. Sample preparation of synthetic fibres is more complicated and, of the several methods available[4], the final choice will depend on the nature of the fibre, and the individual operator. The more suitable methods of sample preparation are described in detail.

Pressed-disc Technique[4]

The two main physical factors in the infrared spectroscopy of solids examined as particles in a non-absorbing medium are the difference in refractive index between the sample and medium and the dimensions of the absorbing particles[5]. In the pressed-disc technique, one can obtain spectra of relatively large particles that are suitable for qualitative identification purposes by choosing as the matrix a halide whose refractive index closely matches that of the sample. In general, potassium bromide (n_D 1·56) is suitable.

Briefly, the method consists of mixing the finely divided fibre with finely powdered potassium bromide which is passed through a 120-mesh sieve and is retained on 200-mesh, dried at 400 °C for not less than 2 hours before use, and should, after drying, be stored in a stoppered bottle in a desiccator.

In preparing the disc, a few milligrams of the fibre are cut up finely with scissors and the process is repeated three or four times, the fibres being rolled into a small ball between each cutting. Some workers recommend grinding the finely cut fibres in a vibratory ball mill but, with those synthetic fibres that are elastic, reduction in particle size is difficult to achieve by this method. A portion of the finely chopped or powdered material is uniformly mixed in an agate mortar with 300–500 mg of finely powdered potassium bromide and pressed into a small disc about 1 mm thick in a suitable vacuum die under a pressure of about 516,750 kN/m². Vacuum alone is applied to the die for 2 minutes, then vacuum and press load are applied simultaneously for 2 minutes. Clear pellets result which have only small absorption bands at 2·9 and 6·1 μm owing to moisture.

It should always be borne in mind that potassium bromide is very hygroscopic and that water-absorption bands, which may be present in spectra run by this method, can lead to wrong identity. Another snag is unequal distribution of fibre in the medium but, for qualitative work, this factor is not important. The potassium bromide method has the important advantage over mulling techniques that extremely small samples may be analysed.

Mulling[6]

This type of sample preparation pertains to solids that do not lend themselves to other methods of preparation. The mulling liquid should be non-volatile and as non-absorbing as possible in the 2–15 μm region. Nujol, which is highly purified mineral oil, is the most readily available and generally useful mulling liquid. Absorption bands, due to the oil, occur at 3·4, 6·9, and 7·3 μm. Mulling agents which are free from absorption in the preceding regions are hexachlorobutadiene and perfluorocarbon oil.

The customary method, whereby the substance is ground to a fine powder from which the mull is prepared, is satisfactory for well-defined crystalline materials, but less satisfactory for textile fibres and inapplicable to viscous, plastic, and rubbery substances. The method described below, as well as being applicable to these relatively intractable substances, is very much faster to operate and the mull is prepared in a single operation. In this method the material is rubbed between ground-glass plates, thus enabling a more powerful abrasive action to be obtained.

The grinding plates are prepared from 5 mm glass plate cut to a convenient size. Pairs of these are ground together with 200-mesh carborundum powder until uniformly rough, then rubbed together using a few drops of Nujol as lubricant until no further glass powder is produced. Minute flat areas with sharp cutting edges are formed on the plates.

Textile yarns or fabrics are cut to short lengths, i.e., about 0·5–2 mm and these are mulled a little at a time, more yarn and Nujol being added at intervals. Excellent mulls of the toughest fibres can be obtained in a few minutes. In preparing a mull, the intention is to produce a paste of Vaseline-like consistency. The correct consistency is judged by appearance, by the drag of the grinding plates, and by the disappearance of such tell-tale signs as rats' tails in the mull that indicate that macroscopic particles are still present. Finally, the plates are separated and the mull is transferred to rock salt plates for infrared measurement.

Solvent-cast Films

In general, a solvent-cast film gives a better spectrum than that obtained by dispersing the same fibre in potassium bromide or in a mull. The cast-film method is not as generally applicable as the pressed-disc technique since a suitable solvent must first be selected, and for some fibres there is no suitable solvent. Further requirements are that the solvent must not react with the fibre and it must leave no residue on evaporation.

If films are cast from a solvent onto a smooth glass surface, the films obtained may produce an interference fringe pattern in the spectrum owing to a high degree of parallelism between their front and back surfaces. The fringes may interfere with the identification of the weaker infrared bands, but the difficulty can be obviated by the simple expedient of using a roughened glass surface. One surface of the film will then be irregular and fringes are not produced.

An approximately 5% solution is made by dissolving the fibre in the hot solvent. Sufficient solution to cover an area of about 50 × 25 mm is poured on to a level glass plate whose surface has been roughened with 400–500-mesh carborundum. The temperature of the solution should be well below that at which bubbles form, otherwise holes are left in the film. Most of the solvent is evaporated off at a temperature low enough to avoid bubble formation and, when the film has solidified, it is heated to a higher temperature, preferably in vacuum, to remove the remaining solvent.

The film can usually be peeled from the glass plate after lifting an edge with a razor blade; wetting with water sometimes helps if the film sticks.

Most solvents are completely removed by the heating, but, where any solvent remains, it may be removed by Soxhlet extraction or refluxing. Form–dimethylamide is tenaciously held by acrylic fibres but is completely removed by boiling the film for ½ to 1 hour in water. It is essential with this method of sample preparation that the solvent be completely removed, otherwise, absorption bands, owing to the retention of the latter, will be present in the spectrum of the fibre.

4.4.1.2   Attenuated Total Reflection (A.T.R.)[7,8]

When a beam of infrared radiation enters a prism it will be reflected internally when the angle of incidence at the interface between sample and prism is greater than the critical angle, which is a function of refractive index. On internal reflection all of the energy is reflected. However, the beam appears to penetrate slightly beyond the reflecting surface, and then return. The depth to which the radiation penetrates is a function of the wavelength, the refractive index of both reflector and sample, and the angle of incident radiation. The apparent depth of penetration ranges from a fraction of a wavelength up to several wavelengths.

If a material that selectively absorbs radiation is placed in contact with the reflecting surface, the beam will lose energy at those wavelengths where the material absorbs owing to an interaction with the penetrating beam. This attenuated radiation, when measured and plotted as a function of wavelength by a spectrophotometer, will give rise to an absorption spectrum characteristic of the material. In order to obtain internal reflection spectra that are nearly identical with transmission spectra, a reflector of relatively high refractive index should be used.

The energy exchange produced by a single reflection is relatively small so that the absorption band produced is weak. Hence the early A.T.R. devices that made use of a single reflection and a variable angle of incidence to produce the desired absorbance often gave weak or highly distorted spectra. In modern equipment the use of multiple internal reflections increases the absorption and produces undistorted spectra of any desired intensity provided there are enough reflections. (See Diag. 4.10.)

Fibres and fabrics, which are among the most difficult materials to handle by transmission spectroscopy, have proved to be quite amenable to study by multiple internal reflection spectroscopy since they require no special preparation techniques for the purpose. The word 'multiple' should be emphasized since the nature of the fibre itself results in poor contact and many reflections are needed in order to ensure sufficient absorption.

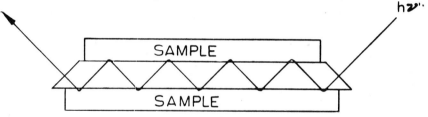

Diag. 4.10 Multiple internal reflection effect

4.4.1.3   Interpretation of Spectra

The infrared method depends primarily upon establishing that the spectrum of the unknown matches exactly the spectrum of a known substance examined in the same physical form, and, in order to do this, it is necessary to be able to name a compound from the absorption bands it displays in the infrared region. In some cases, the spectroscopist may be able to match the unknown with a spectrum from his own files, and for this reason a laboratory carrying out qualitative analyses customarily sets up its own collection of absorption curves of substances it is likely to encounter. Spectra recorded on the same instrument are to be preferred to literature spectra, because no allowance need be made for differences of resolving power or wavelength calibration. When the spectrum cannot be matched in this way, there is still a possibility that a matching spectrum exists in the literature[9].

Since the amount of material in the literature is so great, it is almost impossible to refer to it quickly without some form of mechanical sorting system. Two commercial systems that record infrared data on punched cards are the D.M.S.[10] system and the Wyandotte–A.S.T.M. index[11].

The D.M.S. system contains several thousand cards on which infrared spectra are reproduced; additons are made at regular intervals. These are edge-punched cards which are sorted either on hand needles or with a hand sorting box. Adjacent holes round the edge of the card are assigned to adjacent frequency intervals to cover the spectrum from 2·7 to 40 μm. If the spectrum of the substance to which the card refers contains a strong band in a particular frequency range, the corresponding hole is slotted through to the edge of the card. Other holes on the card are used to denote structural features or properties of the substance.

In order to find the card corresponding to the unknown sample, the frequencies of the strong bands in the unknown spectrum are recorded. Needles are inserted through the holes in the cards corresponding to these frequencies; those cards in which these holes are slotted fall out from the card deck.

The largest index available consists of punched cards issued by the American Society for Testing Materials, which are intended for use on an electrical sorting machine. In order to use the system, the positions of the important absorption bands in the unknown spectrum are estimated to the nearest 0·1 μm in wavelength between 0·7 and 15 μm (or to the nearest μm between 15 and 50 μm). By sorting on the machine, those cards that are punched for absorption bands at the required wavelengths are found. If several cards that bear the same band combination as the unknown are sorted, negative sorting for the absence of absorption bands in particular regions is also possible. The spectra corresponding to the selected cards are found by reference to the identification columns on the cards. These give either the number of the spectrum in one of the abstract compilations, or a reference to the original literature. Several compilations of spectra are indexed in this system.

An extremely rapid system of searching an infrared data file has been described[12]. This method uses a low-cost computer that has been programmed to translate the A.S.T.M. deck of 90,000 data cards into a disc file and to search that file from data entered through the keyboard.

The total search time, excluding data entry and disc warm-up, is approximately 90 seconds. Data entry has been programmed so that the spectroscopist may use the technique himself, after he has received basic instructions. In addition, the form of data entry may be conveniently altered if it is necessary to search the file again.

4.4.1.4 References

[1] C. N. R. Rao. 'Chemical Applications of Infrared Spectroscopy', Academic Press, 1963.
[2] A. D. Cross. 'Introduction to Practical Infra-red Spectroscopy', Butterworths Scientific Publications, London, 1960.
[3] R. W. Hannah and J. S. Swinehart. 'Experiments in Techniques of Infrared Spectroscopy', Perkin–Elmer Corporation, Connecticut, 1968.
[4] Official Digest, Part 2, March, 1961. Federation of Societies for Paint Technology, Philadelphia, U.S.A.
[5] G. Duyckaerts. *Analyst*, 1959, **84**, 201.
[6] A. Crook and P. J. Taylor. *Chem. & Industr.*, 1958, 25 Jan., 95.
[7] B. H. Baxter and N. A. Puttnam. *Nature*, 1965, **207**, 288.
[8] P. A. Wilks and M. R. Izard. 'Internal Reflectance Spectra of Fibres and Fabrics', Wilks Scientific Corporation, 1964.
[9] D. Hummel. 'Kunstoff, Lack- und Gummi Analyse', Carl Hansen Verlag, Munich, 1958.
[10] 'The Documentation of Molecular Spectroscopy (D.M.S.) Punched Cards', Butterworths Scientific Publications, London.
[11] 'The Wyandotte–A.S.T.M. System', American Society for Testing Materials.
[12] D. S. Earley. *Analyt. Chem.*, 1968, **40**, 894.

4.4.1.5 Examples

The spectra were made using a Perkin–Elmer Model 137 Spectrophotometer, and were prepared by the potassium bromide pressed-disc technique.

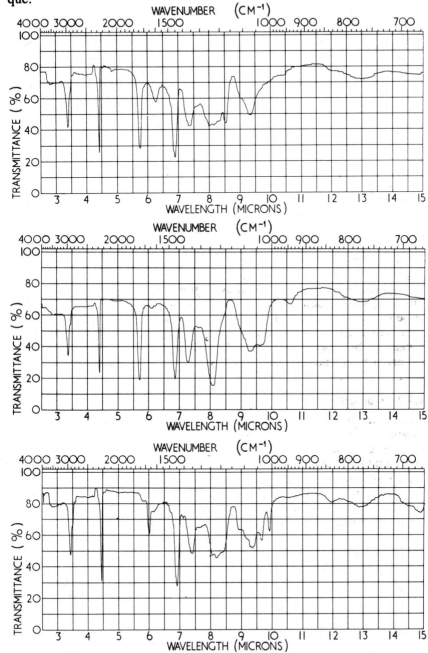

(*a*) Courtelle; (*b*) Acrilan 16, Vonnel 17, ACSA–16[21]; (*c*) Orlon 21

[21]Where more than one fibre trade name is shown the spectra were found to be indistinguishable.

Notes The spectrum of a commercial acrylic fibre consists of a number of bands characteristic of polyacrylonitrile itself, together with other bands characteristic of the minor component of the polymer. Bands common to all acrylic fibres can therefore be readily identified by reference to the spectrum of pure polyacrylonitrile.

Identification of a fibre depends on matching the additional bands with those in the spectrum of a fibre whose identity is already known. In this matching operation both the wavelength of the band peak and its intensity relative to that of a band common to the spectra of all the fibres, e.g., the nitrile band at 4·45 μm, are important. The width of the band, particularly at long wavelengths, is partly dependent on the type of instrument used, and whether it has a linear or non-linear wavelength scale, and is therefore of secondary importance.

On this basis it is possible to represent fibre spectra in a diagrammatic form. In these diagrams the positions of the additional bands present in the spectra of the fibres, but not in that of polyacrylonitrile itself, are shown by vertical lines. The lengths of these lines correspond approximately to the peak intensities of the bands when samples of similar concentration or thickness are examined. An unknown fibre may be identified by noting the positions and intensities of bands not present in the spectrum of poly-acrylonitrile and comparing them with the positions of the vertical lines in the diagrams.

This method has the advantage that the only spectra required are those of polyacrylonitrile, and the unknown fibre itself, but it has the disadvantage that some 'additional' bands can escape recognition when they are close to, or coincident with one of the acrylonitrile bands.

A more satisfactory method is to record the spectra of a comprehensive range of fibres on the operator's own instrument and to compare the spectrum of an unknown fibre with that of each of the known fibres in turn, until an identical one is found.

4.4.2 Gas Chromatography

The technique of gas chromatography was introduced by James and Martin[1] in 1952, and is a special form of the general chromatography procedure of separating a mixture into its components. Separation is achieved by the distribution of the components, as vapours, between two separating media, the stationary one being an involatile liquid known as the stationary liquid phase supported as a thin film on a column of an inert solid support such as a diatomaceous earth, the moving one being a gas that percolates through the column.

The apparatus[2] consists of a source of a suitable gaseous mobile phase such as nitrogen, helium, or argon, the column, and a device for injecting the mixture onto the column either as a gas or as a liquid that is volatilized in a prechamber prior to the column. Additionally the gas stream at the column exit is monitored by a suitable detector to indicate when the sep-arated components emerge. This is usually carried out continuously so that a chart record of some physical or ionization property[2] of the gas stream against time is obtained.

The introduction of commercial instruments for gas chromatography and the development of extremely sensitive detectors has made the technique a very widely applied and sensitive method for the separation and identi-fication of the components of a volatile mixture.

4.4.2.1 Application to Fibres

The criteria for a given mixture to be studied by gas chromatography are that it is stable at the column temperature and sufficiently volatile. A vapour pressure not less than 3 mm mercury is considered desirable[2]. Polymers and resins generally do not satisfy these criteria and cannot be studied directly. However, if a polymer is very rapidly pyrolysed in an inert atmosphere it breaks down to give, among other fragments, a series of volatile products that are then capable of analysis. If the conditions of pyrolysis are sufficiently accurately controlled, the volatile pyrolysis products characteristic of the polymer are reproducibly obtained, and when separated produce a chromatogram characteristic of the polymer[2]. This pyrolysis chromatogram or 'pyrogram' may be used as a 'fingerprint' to help in the identification of polymers and fibres.

4.4.2.2 Procedure

Pyrolysis

The simplest and most favoured type of flash pyrolysis unit incorporates a coil of thin wire suspended in the chromatographic column directly above the packing; the sample is loaded onto the wire, which is then heated by passing electric current through it for about 10 seconds or by induction[3]. About 0·1 mg of fibre can be loaded onto the pyrolysis wire by wrapping a short length around it. Although the pyrolysis conditions are not accurately known they are reproducible and reproducible 'pyrograms' can be obtained for a given sample.

Other designs of pyrolysis unit have been described, giving accurately known and more reproducible pyrolysis conditions than the hot-wire. These are essentially micro-reactor or micro-oven[2] pyrolysis units. In such units the sample (about 0·1 mg) is pyrolysed in a preheated and uniform hot region at about 700 °C. These are probably more suited to identification procedures than the hot-wire. The hot-wire flash pyrolysis technique has been compared to the micro-oven technique[2].

The carrier gas serves as the inert atmosphere for pyrolysis and sweeps the volatile pyrolysis products onto the chromatographic column for separation.

Chromatography

A general purpose chromatographic column may have silicone oil, dinonyl phthalate, or squalane as the stationary liquid phase; more selective phases include polyesters and polyglycols[2,4]. The actual choice of stationary liquid phase and of column temperature depends on the components to be separated and the separation required. For general purposes a silicone oil column at 50 °C or 100 °C would be suitable. However, under flash pyrolysis conditions the products have a very wide range of volatility so that under isothermal conditions of analysis the most volatile components are eluted too quickly to be fully separated and the least volatile are eluted too slowly. Temperature-controlled gas chromatography[2] in which the temperature of the column is increased at a given rate throughout a separation, would give a better resolution of components. The recommended programme depends upon the nature of the stationary phase and the separation required.

4.4.2.3   Applications and Use of Pyrograms

The theory and practice of gas chromatography is described in detail else-where[2,5]. For identification of fibre and fibre-forming polymer, the characteristic 'pyrogram' or pattern of volatile pyrolysis products given off can be used as a 'fingerprint' in a manner similar to that of the infrared spectrum[6]. Some typical pyrograms are shown in the following pages.

Under hot-wire flash pyrolysis conditions, the pyrogram of a fibre is usually complex, so that it would be impracticable to identify the fibre absolutely without some previous knowledge of it. In general, it is better to use other methods to determine the type of fibre and subsequently use gas chromatography as a confirmatory test within a group. This method can be particularly useful for distinguishing one type of dyed acrylic fibre from another. A library of pyrograms is useful, but because of their complexity and their dependence on day-to-day conditions, it is preferable to check the pyrogram against those of authentic samples run on the same day.

4.4.2.4   References

[1]A. T. James and A. J. P. Martin. *Biochem. J.,* 1952, **50,** 679.
[2]W. E. Harris and H. W. Habgood. 'Programmed Temperature Gas Chromatography', John Wiley & Sons Inc., 1966.
[3]W. Simon and H. Giacobbo. *Angew. Chem.* (Internat. Edn), 1965, **4,** 938.
[4]F. Sonntag. *Fette, Seife, Anstrichem,* 1968, **70,** 256.
[5]L. S. Ettre and A. Zlatkis. 'Practice of Gas Chromatography', Interscience-Wiley, 1967.
[6]V. Golscen and D. M. Oates. *Appl. Polymer Symp.,* 1966, **2,** 15.

4.4.2.5   Examples

Pyrograms were obtained under the following conditions:

Apparatus: Philips 4000 series chromatograph and pyrolysis attachment.
     A flame ionization detector was used throughout.

Pyrolysis
 conditions: Specimens were pyrolysed at 770 °C for 2 seconds.

Chromatography: 1. 2 m × 2 mm internal diameter column.
      2. Nitrogen carrier gas at a flow rate of 60 ml per minute.
      3. The stationary phase and programme are given under each pyrogram.

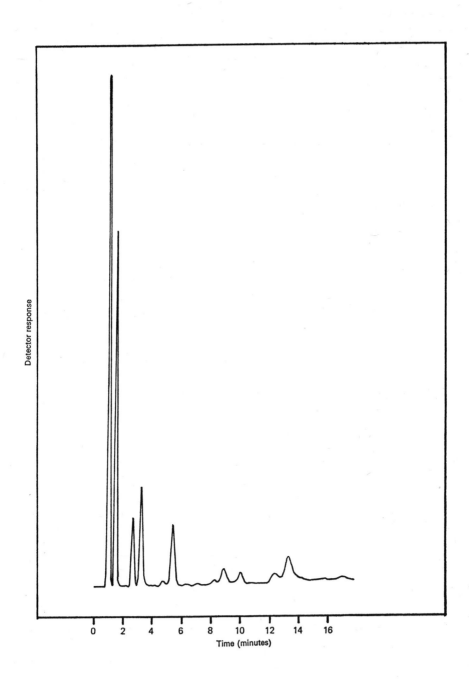

Nylon 6
Stationary phase: 25% dinonyl phthalate on Celite.
Programme: Isothermal at 30 °C for 4 min; then heated to 80 °C at 5°/min.

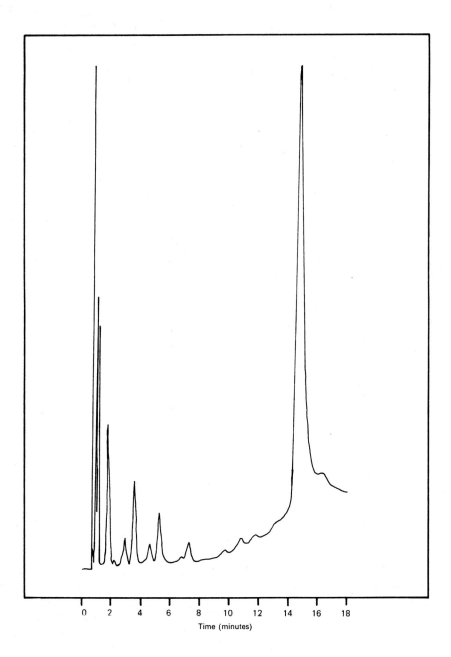

Time (minutes)

Nylon 6.6
Stationary phase: 25% dinonyl phthalate on Celite.
Programme: Isothermal at 30 °C for 4 min; then heated to 80 °C at 5°/min.

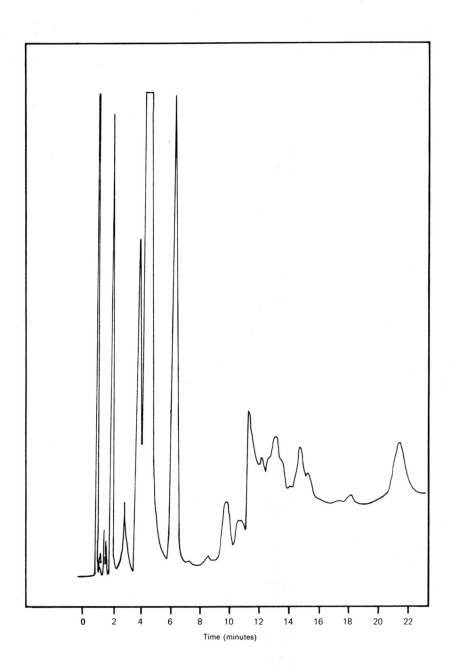

Acrilan 16
Stationary phase: 25% dinonyl phthalate on Celite.
Programme: 30–180 °C at 5°/min.

o

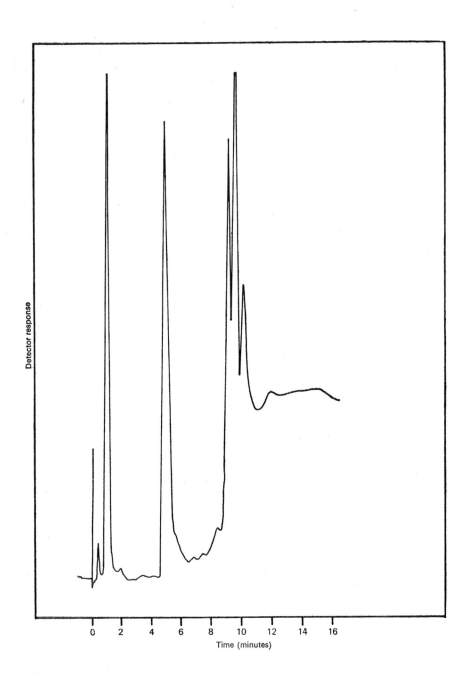

Acrilan 16
Stationary phase: 10% neopentyl glycol sebacate on Chromosorb G.
Programme: 30–180 °C at 15°/min.

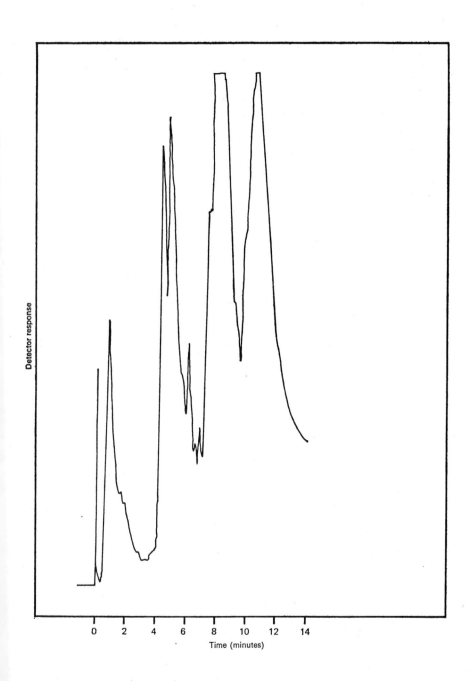

Acrilan 16
Stationary phase: 5% silicone gum rubber SE30 on Celite.
Programme: 30–250 °C at 20°/min.

4.4.3   Differential Thermal Analysis[22]

When a substance is heated, various phase transitions and chemical reactions, involving either the absorption or the evolution of heat, may occur. With fibrous materials, these changes may include the second-order or glass transition, desorption of moisture, crystallization, fusion, chemical reactions, and irreversible decomposition processes. Differential thermal analysis (DTA) is a method whereby enthalpy and heat capacity changes that accompany physical and chemical transformations of matter, may be detected and determined by measuring the temperature differences arising between a sample and a thermostable, inert reference material such as calcined alumina or fine glass beads, as both substances are simultaneously heated or cooled at a predetermined, constant rate of change of temperature.

The differential temperature, ΔT, is measured as a function of either sample, heating block, or reference temperature by means of a pair of series-connected thermocouples or thermistors, one in the sample and the other in the reference material. When the sample is not undergoing any change or reaction there is no potential difference between the two thermocouple junctions since both materials are at the same temperature and therefore ΔT is zero. Whenever a transformation does occur with heat absorbed or evolved by the sample, a temperature difference results, which becomes zero again when the transformation is completed. In practice, ΔT is automatically recorded as a function of temperature or time, to yield a curve or thermogram characterized by the appearance of peaks representing variations from the thermally steady state and corresponding to the various transformations that the sample has undergone. The nature of the measurements makes it possible to distinguish endothermic and exothermic changes; endotherms are usually plotted in a downward direction with exotherms shown as upward deflections as indicated in the hypothetical DTA curve shown in Graph 1, p. 206.

Another more recent differential thermal method that provides similar information involving a different principle, is differential enthalpic analysis (DEA)[1-3]. In DEA, the differences in energy required to maintain the same temperature in both sample and reference during heating and cooling are measured in millijoules/sec as a function of time and/or temperature, whereas in DTA the temperature differences between sample and reference are measured. Thus, when an endothermic transformation occurs, the energy absorbed by the sample is instantaneously replenished by increased energy input with the result that the sample temperature is kept the same as that of the reference. This energy input is recorded, and, since it is precisely equivalent to the energy absorbed, the peak obtained accurately reflects the transition involved. The peak area then provides direct calorimetric measurement of the enthalpy of transition. The data presentation is essentially the same as for DTA.

These dynamic thermal methods permit molecular processes taking place in the solid state to be detected, monitored, and quantitatively measured as they occur. The thermograms of textile fibres reflect molecular and supramolecular structure to provide data and information from a single curve, exemplified by the DTA curve of undrawn Dacron polyester

---

[22]Contributed by R. F. Schwenker, Jun., Personal Products Company, Division of Johnson & Johnson, Milltown, New Jersey, USA.

fibre appearing in Graph 3. Comprehensive treatments of the theory, practice, and instrumentation for DTA are available in recent books and reviews[4-13]. DTA, originally developed in 1889[14], and DEA, which was developed much later[1-3], have been used for the analysis of minerals, metals, and other inorganic materials and, recently, to study polymeric materials. Application to the identification of textile materials is still comparatively new, but more information is becoming available[9,11,13,15-24]. It has been shown that for textile materials, as well as for other substances, the endothermic and exothermic changes observed are both reproducible and uniquely characteristic for a given material, so that the DTA curve or thermal spectrum constitutes a 'fingerprint' and may be used for identification just as an infrared spectrum is so used[8,18,25]. Therefore, the identification of both natural and synthetic textile materials can be accomplished quickly and easily with only a few milligrams of sample. It has also been shown that DTA can detect a physical mixture of polymers[13], so that, in many instances, fibre blends may be characterized by the components present[23]. In Graphs 2, 3, and 4, representative thermograms for various textile materials are shown[18,22,24].

4.4.3.1 Procedure

Modern instruments are now available which are relatively simple to operate and provide useful thermograms of fibrous materials. Descriptions of most current instruments are available in the literature[3-7]. Of prime importance is a temperature programmer or device capable of heating or cooling the samples at a pre-set, relatively constant, rate. The sample and reference material are placed in separate wells in a suitable medium for transferring heat to the thermocouples and samples, usually a metal block or enclosure that is electrically heated, and in which the atmosphere may be controlled. Differential thermal methods are still more demanding than most instrumental methods and the precision and accuracy of results can be quite operator-dependent. The curve or thermogram is affected by sample size, sample geometry, packing, heating rate, atmosphere, signal amplification, thermocouple placement, and reference material. However, with careful attention to experimental variables and an understanding of their effects reliable data can quickly be obtained.

Temperature Range

Most of the available commercial DTA instruments offer a potential operating range of from -150 to $1000\,°C$. In the only available commercial DEA instrument[3] the range is from -150 to $500\,°C$. The temperature range selected depends somewhat on the fibre type and fibre history, but, in general, the range from about $25\,°C$ to about $600\,°C$ is most useful. For most synthetic fibres the DTA curve through the fusion point would suffice for identification, whereas, for the natural fibres and man-made cellulosic fibres, identification depends on recording the high-temperature chemical and decomposition reactions.

Heating Rate

Generally, rates of 5–10 $°C/min$ are satisfactory and permit a useful thermogram to be obtained, but linear temperature programming from

0·5 °C to 80 °C/min is available. Characterization studies on several different synthetic and natural fibres have shown that well-defined accurate thermograms can be produced in less than an hour at a rate of 10 °C/min[18-24].

Fast heating rates tend to yield sharp major peaks, but if the rate is too fast detail is lost and minor reactions are obscured. Conversely, slow rates of less than 5 °C/min permit small changes to be detected and facilitate reaction separation, but make peak area more difficult to define. Thus, the selection of heating rate is always a compromise. In general, reversible transition temperatures, e.g., melting, are essentially independent of heating rate whereas irreversible reaction peak temperatures are dependent on heating rate. In the latter case, peak temperatures are moved upward by increasing the heating rate. Heating rates need to be carefully specified so that invalid comparisons are avoided.

Control of Atmosphere

Atmosphere is an important variable in the thermal degradation of polymeric materials[6, 7, 13, 14]. To prevent hard-to-control oxidative reactions at elevated temperatures that often produce poorly defined curves, it is recommended that an inert atmosphere, preferably dry nitrogen, be used[23].

Sample Considerations

Sample size and disposition are controlled, in part, by the sample-holder design and instrument sensitivity. Sample-holders vary from relatively massive metal blocks containing wells of varying diameter, capable of containing as much as 1 g or more of sample, to isolated, shallow pans where sample sizes down to 0·1 mg may be used. The instruments commercially available permit sample sizes from less than 1 mg to 50 mg to be used.

In general, the smaller the sample the better the results. Since a finite time is required for a transformation to occur, a large sample could affect the observed temperature[26] and give rise to a substantial thermal gradient within the sample[6, 8]. In a small sample (1–10 mg) the thermal gradient is reduced, only semi-micro quantities are required, and volatile degradation products are more readily released. However, care must be exercised since for highly heterogeneous systems a very small sample may not be representative, leading to poor curve reproducibility[6]. Instrument sensitivity and electronic 'noise' are also controlling factors, since very small samples can only be used when high amplification of the signal is possible with low 'noise' level.

The fibre must be finely divided and excellent results can be obtained when fibre, yarn, or fabric samples are cut into 2–3 mm lengths or into small squares[18, 22, 23]. If low-temperature transitions are sought, it is not good practice to grind samples to a powder, because, in the process, sufficient heat can be generated to change the sample history[22, 24]. However, where only high-temperature changes are of interest, polymer and fibre samples may be ground or milled. In general, it is the usual practice to run the sample as received and if size reduction is required, the most gentle process is used to accomplish the purpose.

With respect to sample packing, the techniques are as follows:

(*a*) unadulterated sample;

(*b*) 'sandwich' packing wherein a compressed pellet of ground sample is placed between layers of reference materials[4, 16]; or

    (*c*) admixing the sample with reference material as a diluent to yield sample concentration of 5–30%.

It is advantageous to dilute relatively large samples, greater than 10 mg, with the inert reference material so that differences in heat capacity, thermal conductivity, and volume changes will be minimized. These changes will then be principally determined by the reference material regardless of sample shrinkage, melting, decomposition, etc.[4, 9–13, 25]. A diluent also effectively reduces thermal gradients within the sample by separation of the total into smaller aggregates.

As a result of studies on textile fibres[22, 23, 27], a modified 'sandwich' packing was developed wherein one-third of the total amount of reference material to be used is placed in the sample well as a base layer. The sample is then mixed with about one-third of the reference material, and the mixture so disposed that approximately half of the sample is below the thermocouple bead and half above the thermocouple bead. The final third of the reference material is then deposited as a top layer.

Reference Materials

The reference material should be inert and thermostable over the temperature range used. If a diluent is used or required, it should be the same as the reference. The most commonly used reference material is calcined alumina, aluminium oxide (Al_2O_3). Silica, silicon carbide, glass beads, glass wool, and quartz wool have also been used[4–6]. In most instances where sample size is 0·1 to 10 mg, no reference material is needed other than the empty pan on the reference side.

4.4.3.2 *Identification*

The DTA curve of a synthetic material provides much more than an identifiable curve, since acceptable data on second-order transition, crystallization, and melting point are also obtained as shown in Graph 4. Identification of synthetic fibres can be achieved by (i) comparing the curve of the unknown with published curves[8, 9–13, 15–24] (ii) running an authentic sample under identical conditions for comparison, and (iii) checking the melting point and other transition data.

4.4.3.3 *References*

[1] L. M. Clarebrough, M. E. Hargreaves, D. Mitchell, and G. W. West. *Proc. Roy. Soc.,* 1952, **A215,** 507.

[2] C. Eyraud. *Comptes Rendus,* 1954, **238,** Part 2, 507.

[3] E. S. Watson, M. J. O'Neill, J. Justin, and N. Brenner. *Analyt. Chem.,* 1964, **36,** 1233.

[4] W. J. Smothers and Y. Chiang. 'Differential Thermal Analysis', Chem. Publ. Co. Inc., New York, 2nd edition, 1966.

[5] W. W. Wendtlandt. 'Thermal Methods of Analysis', Interscience, New York, 1964.

[6] P. D. Garn. 'Thermoanalytical Methods of Investigation', Academic Press, New York, 1965.

[7] P. E. Slade, Jun., and L. T. Jenkins. 'Techniques and Methods of Polymer Evaluation', Marcel Dekker Inc., New York, Vol. I, 1966.

[8] R. F. Schwenker, Jun. 'Analytical Methods for Textile Laboratory', AATCC Monograph, (edited by J. W. Weaver), American Association of Textile Chemists and Colorists, Research Triangle Park, North Carolina, 1968, Chapter IX.

[9] R. C. MacKenzie and B. D. Mitchell. *Analyst,* 1962, **87,** 420.

[10]B. Ke (ed.). 'Thermal Analysis of High Polymers', Polymer Symp. No. 6, Interscience, New York, 1964.

[11]R. F. Schwenker (ed.). 'Thermoanalysis of Fibers and Fiber-forming Polymers', Appl. Polymer Symp., No. 2, Interscience, New York, 1966.

[12]C. B. Murphy. *Analyt. Chem. (Analyt. Rev.)*, 1966, **38**, 443R; *Analyt. Chem.*, 1968, **40**, 380R.

[13]B. Ke. 'Differential Thermal Analysis', *in* Encyclopedia of Polymer Science and Technology, Interscience, New York, Vol. 5, 1966.

[14]W. C. Roberts-Austen. *Proc. Instn Mech. Engrs,* 1899, **1**, 35.

[15]D. Costa and G. Costa. *La Chimica e L' Industria,* 1951, **33**, 71.

[16]H. Morita and H. M. Rice. *Analyt. Chem.,* 1955, **27**, 336.

[17]T. R. White. *Nature,* 1955, **175**, 895.

[18]R. F. Schwenker, Jun., and L. R. Beck, Jun. *Text. Res. J.,* 1960, **30**, 624.

[19]R. F. Schwenker, Jun., and J. H. Dusenbury. *Text. Res. J.,* 1960, **30**, 800.

[20]B. Ke and A. W. Sisko. *J. Polymer Sci.,* 1960, **42**, 15.

[21]J. J. Keavney and E. C. Eberlin. *J. Appl. Polymer Sci.,* 1960, **3**, 47.

[22]R. F. Schwenker, Jun., and R. K. Zuccarrello. *J. Polymer Sci.,* 1964, **C6**, 1.

[23]R. F. Schwenker, Jun., L. R. Beck, Jun., and R. K. Zuccarrello. *Amer. Dyest. Rep.,* 1964, **53**, 30.

[24]R. F. Schwenker, Jun. 'Proceedings of the Second Toronto Symposium on Thermal Analysis', (edited by H. G. McAdie), Chemical Institute of Canada, Toronto, 1967.

[25]C. B. Murphy. *Mod. Plastics,* 1960, **37**, 125.

[26]D. A. Vassallo and J. C. Harden. *Analyt. Chem.,* 1962, **34**, 132.

[27]R. F. Schwenker, Jun., L. R. Beck, Jun., and W. Kauzmann. Final Report U.S. Navy Contract N140 (132) 5744B, Feb., 1960.

4.4.3.4 Examples

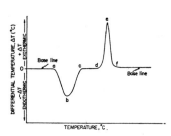

Graph 1. Hypothetical DTA curve (thermogram):
  (i) Peak a b c—endothermic transformation
  (ii) Peak d e f—exothermic transformation
  (iii) Temperatures at b and e—peak temperatures.

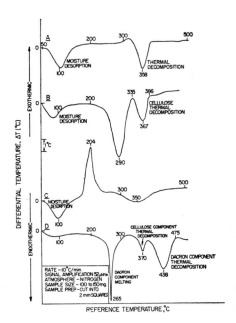

Graph 2. DTA curves of unmodified and modified cotton fabrics:
  A—Purified cotton fabric (100 mg sample)
  B—Cyanoethylated cotton fabric (%N = 5·96, DS = 0·9, 100 mg sample)
  C—Tosylated cotton fabric (DS = 0·25, 150 mg sample)
  D—35% cotton/65%Dacron blend fabric (100 mg sample).

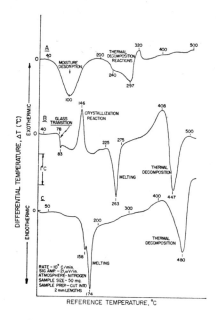

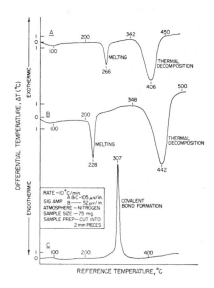

Graph 3. DTA curves of textile fibres:
A—American domestic wool
B—Undrawn Dacron (commercial)
C—Drawn polypropylene (experimental).

Graph 4. DTA curves of textile materials:
A—Nylon 6.6 fabric
B—Nylon 6 fibre
C—Orlon 42 fabric.

4.4.4   Thermogravimetric Analysis

Thermogravimetric Analysis (TGA) or thermogravimetry involves the use of a recording thermobalance to automatically measure and record weight changes which occur when a sample of material is heated or cooled to a controlled programme in a controlled environment. The resulting weight change *vs* temperature curve, or its derivative, gives information concerning the thermal stability and composition of the original sample.

4.4.4.1   References

C. Duval. 'Inorganic Thermogravimetric Analysis', Elsevier Publishing Co., Amsterdam, 1953.
P. D. Garn. 'Thermoanalytical Methods of Investigation', Academic Press, New York, 1965.
W. W. Wendtlandt. 'Thermal Methods of Analysis', Interscience, New York, 1964.

4.4.5   Scanning Electron Microscopy

Thermogravimetric analysis (TGA) or thermogravimetry involves the use textile materials that either could not be seen before or that required long, tedious work on the light microscope and the skill of an experienced microscopist. It is not intended to suggest that an SEM is a necessity for fibre identification, as a light microscope is perfectly adequate and much cheaper, however, if an SEM is available, or if SEM pictures are needed for some other part of an investigation, this can provide a useful aid to identification. Fibres with a characteristic shape or surface structure, e.g., cotton, wool, viscose, trilobal man-made fibres, etc., are easily recognized and their proportion and location in the material can be estimated (Figs 1, 37, 41, 42, 77, 78, 89, 115, 137). The three-dimensional view which is obtained with the SEM, but which is not possible with the light microscope, is illustrated by comparing Figs 40 and 41.

SEM[1] was first made commercially available in 1965, following the original idea for an SEM in 1935 and the significant improvements in design made by C. W. Oatley and his group at Cambridge. The basic method of operation of the SEM is the emissive mode whereby the surface topography of the specimen is examined. The image seen on the viewing screen is an indirect electron image and it is fortunate that this image is similar to that which one would expect to see under normal lighting conditions. The advantages of the SEM over light microscopy are the increased depth of focus, greater resolution, increased ease of specimen manipulation, greater range of magnification ($\times 20-\times 100,000$), and the ability to examine the specimen from many different aspects. The main disadvantages are the high capital cost of equipment and cost of upkeep, high cost of ancillary equipment, the fact that the specimen is examined in a vacuum environment, and that as textile materials are poor conductors of electrical charges, they have to be coated with an electrically-conducting material.

Over the past ten years other modes of operation have been developed for the SEM; these include cathodoluminescence, specimen current amplification, and x-ray emission. The various modes are used for the examination of many different types of material, but the most useful for textile materials is still the emissive mode. The SEM can be used to examine:

(*a*) dimensional shape of fibres from different angles,
(*b*) surface detail of fibres,
(*c*) modifications in fibre shape or surface detail,
(*d*) fibre damage,
(*e*) fibre distribution in yarns and fabrics,
(*f*) yarn and fabric construction,
(*g*) fractography of fibres broken by various means, and
(*h*) nonwovens.

The SEM cannot be used to examine internal detail and colour, owing to the nature of the electron image.

The specimen is rapidly scanned by a beam of electrons produced by a heated tungsten filament. The electrons travel from the gun through an evacuated column passing through electromagnetic lenses which focus and reduce the diameter or spot size of the beam. The beam of electrons is scanned across the specimen by the scanning coils. High energy and low energy (secondary) electrons are emitted from the surface of the specimen and these are attracted to the collector cage and impinge onto the scintillator tip of the light guide. They are converted to photons of light which pass down the light guide to the photomultiplier, which multiplies them through various stages, and they are then converted back into an electronic signal by the head amplifier. This signal is produced on the cathode ray viewing screen; the image is built up by the spot scanning the screen in synchrony with the electron beam on the specimen. If the SEM is equipped with TV scanning coils, the image can be displayed on a TV monitor. This makes is much easier to manipulate the specimen, which is particularly useful when the surface of the specimen has to be methodically searched.

4.4.5.1 Specimen Preparation

As textile materials are poor conductors of electrical charges, static charges can build up. It is important, therefore, that the sample is correctly mounted and prepared so as to eliminate, wherever possible, any charging effects. Charging gives poor picture quality, image distortion, and can introduce spurious artifacts that can give misleading results.

Fibres, yarns, and fabrics can be mounted directly on to flat SEM specimen stubs using a liquid adhesive or double-sided adhesive tape. Liquid adhesives should be used with care as the specimen can sink into the adhesive and be partially covered; it is recommended that most textile samples are mounted on conducting adhesive tape. The procedure for mounting single fibres is relatively simple but great care has to be taken otherwise a lot of effort is wasted. It is recommended that a fibre or fabric specimen is scoured before mounting on a specimen stub to remove any finishes which could obscure surface characteristics, but this will depend on the reason for the SEM examination.

Separate a yarn threadline into its individual fibres and randomly select about twelve fibres to be mounted on an SEM specimen stub. Mount a stub in a holder (one can easily be made from Meccano) and run two parallel strips of adhesive, such as Durofix, at opposite sides leaving a clear area in the stub centre. Using two pairs of tweezers and a bench magnifier, carefully place individual fibres at evenly spaced intervals across the stub from one run of adhesive to the other (Fig. 160). Sufficient tension must be applied in the mounting to keep the fibres straight, unless bulk is being examined, when only a minimum of tension should be applied. When the adhesive has set, the stub is ready to be vacuum coated.

This method has the advantage of leaving a reasonable area of fibre free from any adhesive material, thereby permitting a comparatively large area of surface to be examined. The same procedure is used in the mounting of whole yarn threadlines, but for fabrics it is better to use double-sided adhesive tape over the whole of the stub surface and to stick the fabric to this. The overlapping edges are then trimmed off with scissors to the stub edge and adhesive run round the outer edge (Fig. 161). The samples are then vacuum coated with a conducting metal in the usual way.

Fibre Cross-sections in the SEM

Fibre cross-sections can be examined in the SEM using a similar technique to that for light microscopy. A stub can be made which is based on a tongue and slot, with the tongue half capable of being clamped to the half with the slot in order to hold the whole stub together.

The two halves of the stub are separated and the fibres placed in the slot. The two halves of the stub are clamped together, sufficiently tightly to hold the fibres without crushing them. The surplus fibres above the stub surface are cut level with the surface using a sharp, single-edged razor blade. The stub is now ready to be vacuum coated in the usual manner (Fig. 158).

This style of stub which can be separated into halves can also be used for the SEM examination of broken fibre ends and obviates the use of adhesive tapes in a high vacuum system (Fig. 159).

If fibre ends are to be examined, they should be held in a vertical position in a specially adapted split stub[2] so that the ends can be viewed from all directions. The split stub has a thicker base than normal, from which a step is cut. The step portion covers half the surface area of the stub and the two portions are held together by a screw. It is not possible, except with the coarser monofilaments, to mount fibres directly into the stub. The fibres are first mounted on conducting adhesive tape so that their ends project from the edge and the tape is folded over so that the fibres are sandwiched between the adhesive layers. The portion containing the fibre ends is cut out and mounted in the split stub, the two halves of the stub are pushed together to trap the tape, and are securely screwed together. The fibres are now held in an upright position at the centre of the stub. The fibres are then coated with a conducting layer of metal. It is recommended that with very fine diameter fibres not more than 1 mm of fibre is allowed to project from the edge of the tape, otherwise there is a danger of fibre movement when the specimen is bombarded with electrons. There are two methods of making textiles conducting by vacuum deposition of a conducting metal onto the surface of the specimen.

Metal Coating by Thermal Evaporation

Specimens are usually vacuum coated with a metal; silver, gold, or gold–palladium are used in commercially available equipment. Atoms of metal emitted from an evaporation source travel in straight lines and are only deflected by collision with air molecules and, as the evaporation chamber is evacuated before coating commences, there are few air molecules left to cause deflections. If the specimen is stationary only those surfaces exposed directly to the source will be metal coated. Samples that are only partially coated can charge up in the SEM, but this problem can be corrected by exposing the sample to more than one source or by rotating it, usually at an angle to one or more sources. Such devices can be constructed within the evaporation chamber or planetary movements can be bought commercially.

The thickness or the metal coating must be sufficient to prevent charging, but not so thick as to block out surface detail. Metal coating is needed to improve the emission of electrons from the surface of the specimen; if the electron emission is poor, the electron image displayed on the viewing screen is electronically noisy giving poor picture quality and resolution.

Metal Coating by D.C. Sputtering

Another method of metal coating a sample, that has been recently introduced to SEM work, is D.C. sputtering. The mounted samples are placed inside the chamber of the sputter coater and the chamber evacuated to the required pressure. An inert gas, usually argon, is bled into the chamber at a set flow rate and voltage is applied to the anode which ionizes the gas atoms making them positively charged. These positively charged atoms crash into the cathode (target) displacing metal atoms from its surface. The displaced atoms descend as a cloud completely covering the specimens below. The cathode can be made of gold, silver, or other metals, or can be a base metal coated or plated with gold or silver. This method of coating takes longer to complete than evaporation coating but is more efficient.

Sputtering units can be bought commercially or it may be possible to convert the evaporation coating unit in use to D.C. sputtering. The sputtering units commercially available usually coat only with gold because the ultimate vacuum achieved is not low enough for other reactive metals. Where lower pressures can be obtained, other metals can be used for coating.

Antistatic Agents

Antistatic agents can also be used to give a conducting surface; these are solutions of long-chain organic liquids that suppress the formation of static electricity at the surface of some insulators when applied thinly[3]. They can be used in aerosol form and sprayed onto the sample, or dissolved in a solvent to a very dilute solution in which the sample is soaked. Antistatic solutions are especially useful for experiments within the SEM where the specimen cannot be metal coated, e.g., stretching the sample, and for samples which pose problems in metal coating.

4.4.5.2 Examination in the SEM

The prepared specimen is mounted on the specimen stage inside the specimen chamber. The chamber is evacuated and the specimen exposed to the electron beam; the specimen can be moved in the X, Y, and Z directions and can be tilted and rotated with respect to the beam. The accelerating voltage applied to the gun is very important as it has been found that specimens of low density can be damaged by high kV beams and may exhibit charging phenomena. It has also been found that at a high kV there is some loss of image detail owing to the deeper penetration of the beam within the specimen and increased emission at thin edges[4,5]. It is recommended that about 5 kV accelerating potential be used for textile materials which have been evaporation coated with metal. Charging problems are much less for specimens that are sputter-coated with metal and up to 10 kV can be used without too much loss of image detail. However, the operating conditions used depend on the make of SEM, on the sample, and on the skill of the operator.

A record is made on standard photographic film from either a short-persistance phosphor record screen or, as on some SEMs, a combined visual–record screen.

4.4.5.3 References

[1] J. W. S. Hearle, J. T. Sparrow, and P. M. Cross. 'The Use of the Scanning Electron Microscope', Pergamon Press, Oxford, 1972.

[2] P. M. Cross, J. W. S. Hearle, B. Lomas, and J. T. Sparrow. Proceed. 3rd. Ann. Symp. on S.E.M., 1970, p. 81.

[3] J. Sikorski, J. T. Moss, A. Hepworth, and T. Buckley. *J. Sci. Inst. Series 2*, 1968, **1**, 29.

[4] J. W. S. Hearle, B. Lomas, and J. T. Sparrow. *J. Microsc.*, 1970, **92**, 205.

[5] A. Boyde and C. Wood. *J. Microsc.*, 1969, **90**, 221.

5 Notes on Reagents
and
Methods of Test

Preparation of Sample for Analysis

In order to carry out identification tests successfully, dyes, sizes, fillings, and finishes, etc., which might interfere with the tests, should be removed. Staining tests must obviously be affected by heavily dyed material, so it is necessary to remove the dye from such materials before applying any staining test. A variety of reagents is available for stripping dyes but no single reagent will serve to remove all types of dye from all types of material and a careful selection must be made. It is particularly important to realize that many of the stripping agents affect the reactions of certain fibres to staining tests. If, for example, an acetate-fibre fabric were stripped with a hot alkaline solution, staining tests on the stripped material would give misleading results. To avoid such error, it is recommended that fibres of known origin should be subjected to the stripping treatment given to the fibre under test and a record kept of their reactions to the staining technique employed. Loom-state materials may contain size or oil that must be removed before staining. Oil may be removed by extraction with suitable solvents such as petroleum ether.

Brief notes, covering the stripping of dyes, and the identification of finishes are given below. For information on the removal of finishes see Appendix A.

5.1   Stripping of Dyes

Before resorting to the use of powerful solvents or strongly alkaline reducing solutions for the removal of dyes, a reducing treatment under almost neutral conditions should be tried. Most azo dyes including those of the reactive classes are decolorized at the boil.

5.1.1   Reducing Agents

Neutral reduction. Place the fabric or fibres into a warm solution of 10 ml water containing 2 drops 0·880 ammonia and 0·5 g sodium hydrosulphite. Raise the solution to the boil and continue gentle boiling until decolorization is complete. Wash the sample thoroughly in warm water and dry.

Boiling Sodium Hydrosulphite (5%) containing sodium hydroxide (1%) strips many classes of dye and is probably the most generally useful stripping agent, but it is not suitable for animal fibres, or cellulose acetate. The addition of 15% butyl carbitol is helpful in the case of dyes that are difficult to strip.

Sodium Sulphoxylate–formaldehyde (Formosul). A boiling solution containing approximately 2% Formosul and 0·5% acetic acid is suitable for cellulose acetate fabrics.

5.1.2 Solvents

Pyridine. Soxhlet extraction with 20% aqueous solution of pure pyridine will completely remove many direct and disperse dyes.

Form–dimethylamide. Soxhlet extraction strips azoic colours and some vat colours from cellulose. The solvent should be used in a fume cupboard.

Monochlorobenzene can also be used for the removal of disperse dyes; the temperature must not exceed 100 °C with cellulose acetate fibres. Soxhlet extraction removes most disperse dyes from polyester fibres.

Boiling *o*-Chlorophenol strips most vat and azoic dyes. Stripping is facilitated if cellulose fibres are first swollen by boiling for about 1 minute in 10% aqueous urea solution. *o*-Chlorophenol is a solvent for nylon. Great care should betaken in the use of this solvent; skin contact should be avoidcd.

5.1.3 Acids

Boiling 5% Acetic Acid removes basic dyes from silk or wool.

5.1.4 Alkalis

Boiling 1% Ammonia (0·880) removes acid dyes from silk or wool.

5.1.5 Oxidizing Agents

Sodium Hypochlorite (0·1N) at room temperature and at pH 10–11 adjusted with caustic soda, if necessary, is safe for cellulosic fibres. Its use is restricted, however, and better results can generally be obtained with alkaline hydrosulphite.

Sodium Hypochlorite (0·04N) acidified with acetic acid or sulphuric acid, quickly strips many colours at room temperature, but the reagent also attacks cellulose rapidly.

Boiling Sodium Chlorite (2%) adjusted to pH 5 with dilute acetic acid, will strip aniline black, sulphur, and other dyes. This does not tender the fibre. The addition of a little hydrogen peroxide to the stripping solution will reduce the liberation of objectionable fumes of chlorine dioxide.

5.2 Detection of Formaldehyde Treatments of Fabrics

Fabrics may be resin-treated for a variety of reasons, most commonly to produce resistance to creasing, permanent embossed effects, minimum-iron properties, or dimensional stability. Cellulosic fibres from fabrics that have been treated by amino–formaldehyde processes are generally insoluble in cuprammonium hydroxide and the stain obtained from Shirlastain A is different from that of untreated material. The value of staining tests on coloured finished materials is limited by the original colour.

Cross-linking treatments with formaldehyde and glyoxal may also be encountered, again modifying the fibres so that they are insoluble in cuprammonium hydroxide.

While it is not the purpose of this book to identify resin finishes on fabrics it may be necessary to confirm the presence or absence of such treatments, and the following methods are suggested.

5.2.1   Tests for Amino–formaldehyde Resins

Soda-lime Test. See Section 5.16.9. This initial test is to determine the presence or absence of nitrogen.

Furfural Test. Boil about 0·1 g of material with 5 ml of the following solution for 5 min:

> 20 ml acetone
> 20 ml distilled water
> 10 ml conc. hydrochloric acid
> 1 ml freshly re-distilled furfural.

This solution should not be kept longer than two hours. A red colour develops if urea or triazone resins are present. Other resins develop an orange to yellow colour.

Picric Acid Test. Prepare an extract of the finished fabric by boiling for 3 min in 0·1N hydrochloric acid and decant or filter the extract from the sample. Add a few drops of saturated picric acid solution to the acid extract from the fibres. A bulky yellow precipitate is formed in the presence of melamine. Other resin hydrolysates do not precipitate.

5.2.2   Test for Formaldehyde

Chromotropic Acid Test. Scour the fabric in 1 g/l sodium carbonate at 60 °C for two minutes, rinse, and dry. Prepare an acid extract of the scoured material as in the Furfural Test above. Cool and filter the solution. To 1 ml of the solution, add 3 ml conc. sulphuric acid and 1 ml 0·2% chromotropic acid solution, and heat the solution at 60 °C for 10 min. A deep red-violet colour is produced in the presence of formaldehyde.

5.3   Identification of Resin Treatments

5.3.1   Identification of Amino–aldehyde Resin on Textiles by Paper Chromatography

Note. The following is intended only as a general guide to the principles of chromatography. Cloth preparatory[1] and stripping techniques, etc., vary from laboratory to laboratory, and the references given should be consulted.

Procedure. About 2 g of fabric are boiled with 100 ml of 0·1N HCl or H_2SO_4 for approximately 2 hours, maintaining constant volume if necessary, or until only traces of formaldehyde remain (Chromotropic Acid

[1]J. H. Howard. *Amer. Dyest. Rep.*, 1957, **46**, 313.

P

Test). The solution is then neutralized[3] before it is evaporated to 5 ml and the amine hydrolysate (25 to 100 μl) is spotted[1, 2] onto Whatman No. 1 chromatography paper prepared[1, 3, 4] in the usual way. Several repeat spots of known and unknown hydrolysates are included.

The chromatogram can then be eluted in a chromatography tank by ascending or descending techniques[1, 3, 4] using a mixture of n-butanol, ethanol, and water (3 : 1 : 1 by volume) as eluent[1, 3]. After eluting to about 40 cm (for good compound separation) the solvent front is marked and the paper dried before cutting into separate sections for spraying with the appropriate reagents. The positions and identity of the various amino compounds are revealed as coloured spots.

> *Part A.* Spray with a solution obtained by diluting 1 ml saturated aqueous silver nitrate with 200 ml acetone, then add water drop by drop until the precipitate produced is redissolved by stirring.
>
> Next, spray with a solution obtained by diluting 4 ml 50% w/v aqueous sodium hydroxide with 100 ml ethanol. The twice-sprayed paper is allowed to dry, dipped in 1N ammonium hydroxide to clear the brown background, rinsed with water, and dried.
>
> *Part B.* Sprayed with *p*-dimethylamino benzaldehyde in ethanol and hydrochloric acid[5]—detects urea and ethylene urea.
>
> *Part C.* Sprayed with ninhydrin and developed at 100 °C (5 min)[5]—detects amines from triazone resins.
>
> *Part D.* Saturated with chlorine gas and sprayed with *o*-tolidine[2]—detects all amino resin hydrolysates.

Identification is achieved by comparison of the position of known and unknown spots (R_f values) and colour on the developed chromatogram.

More recently, thin-layer chromatography (T.L.C.) has been used for identification[5, 6].

5.3.2   Identification of Silicone Finish on Textiles

Into a *new unused* ½-inch-diameter test tube, place approximately 1-cm depth of conc. sulphuric acid. Add the specimen to be examined (e.g., 1 cm² of fabric) and bring almost to the boil while shaking gently. Water will initially reflux up and down the wall of the test tube, wetting it completely. If a silicone finish is present, it will transfer from the textile to the glass, which will cease to be 'wetted'. This is most easily seen when the test liquor is black. If the specimen does not contain a cellulosic fibre, a small amount should be added to turn the acid black. Do not boil for more than about ½ min, as the silicone is gradually destroyed by the acid.

A *new* tube must be used for each test.

[1] J. H. Howard. *Amer. Dyest. Rep.*, 1957, **46**, 313.

[2] J. C. Brown. *J. Soc. Dyers Col.*, 1964, **80**, 185.

[3] J. C. Brown. *J. Soc. Dyers Col.*, 1960, **76**, 536.

[4] J. G. Feinberg and I. Smith. 'Chromatography and Electrophoresis on Paper', Shandon Scientific Co. Ltd.

[5] L. Meckel, H. Milster, and U. Krause. *Textil-Praxis*, 1961, **16**, 1032.

[6] N. Buchsbaum and A. Datyner. *J. Soc. Dyers Col.*, 1966, **82**, 18.

5.4 Refractive Indices of Fibres

Refractive index governs the visibility of all colourless and transparent objects and is one of the chief considerations in the choice of a mounting medium. Furthermore, it is one of the few numerical constants that can be determined by the microscope with accuracy and it constitutes a useful criterion in the classification and identification of solids and liquids.

When a fibre is examined in air (n 1·0), the relatively large difference in refractive index between the fibre and air causes about 5% of the incident light to be reflected and the transmitted light to be markedly refracted. These effects produce heavy shadows that obscure fine details of the fibre structure and can introduce misleading artifacts. To reduce the degree of contrast in the shadow regions the fibres are mounted in a medium of suitable refractive index.

If fibres are mounted in a medium of similar refractive index, surface characteristics are practically invisible but internal structure and the presence of voids, or inclusions such as pigmentation, are clearly revealed. It is recommended that with fibres that have fairly high positive birefringence values, i.e., where the difference between $n_{\parallel}$ and $n_{\perp}$ is fairly large, that the mountant is matched with $n_{\parallel}$ of the fibre using plane polarized light with the plane of vibration perpendicular to the fibre axis. This is because light transmitted vibrating parallel to the fibre axis (i.e., the extraordinary ray) does not follow the normal laws of light and the image quality is poorer. When it is desired to examine surface detail of the fibres, a mounting medium of significantly different refractive index should be selected, preferably with $n_{\parallel}$ higher than that of the fibre, e.g., 1-bromonaphthalene or di-iodo methane, but mountants of significantly lower refractive index can be used.

Mountants should be relatively stable and involatile liquids that are unreactive with the usual polymers used for fibres. The most commonly used mountant for fibre identification is liquid paraffin which gives an image of satisfactory contrast for all fibres except cellulose diacetate and cellulose triacetate for which n-decane is recommended. Glycerine jelly is satisfactory with natural cellulose fibres, but has a small swelling effect; methyl salicylate is a good match for cotton and viscose but has a pungent smell (oil of wintergreen). Water can be used for examinations to determine fibre appearance, but, because it will swell fibres capable of absorbing moisture, it must not be used as a mountant when measuring fibre diameter.

It is recommended that fibres are examined as soon as possible after mounting; some fibres are likely to be penetrated in time by certain mountants, e.g., polyester fibres are affected by 1-bromonaphthalene. Also some mountants evaporate more quickly than others, e.g., n-decane.

The liquids listed in Table B3.1 (Appendix B) are suitable fibre mountants but some are harmful either by inhalation, skin absorption, or otherwise. Every care should be taken to note and follow the manufacturer's instructions given on the bottle. Values of $n_{\parallel}$ and $n_{\perp}$ for different fibres are given in Table B3.2 (Appendix B) and which should be referred to when choosing a suitable mountant.

In general, fibres vary slightly in refractive index from place to place both axially and transversely. Because of this, it is possible to bring the refractive index of the medium to exact equality with the fibres only locally.

These local variations of refraction are generally detectable only by refined methods and any region of markedly different refraction probably indicates the presence of another substance such as air, oil, or titanium dioxide.

Factors governing the refractive index of fibres are the chemical nature of the molecules, the physical arrangement of these molecules, the wavelength of the incident light, moisture content, and other substances that may be present in the fibres. In order to make accurate determinations it is necessary to use plane-polarized light under controlled conditions of temperature and relative humidity.

Birefringent, i.e., optically anisotropic, substances exhibit different indices of refraction for a given wavelength depending on the direction of vibration of light passing through them, as well as upon its direction of transmission. In positively birefringent fibres the directions of maximum and minimum refractive index correspond to the long axis of the fibres and the direction at right angles to the axis respectively; for negatively birefringent fibres the converse is true. In some natural fibres these directions are parallel and perpendicular to a spiral direction in the fibre.

The values of $n_{\parallel}$ and $n_{\perp}$ given in Table B3.2 were measured on an interference microscope which is the most accurate method available. However, they can be determined fairly precisely in monochromatic light using a range of liquids of known refractive indices. These liquids can be prepared by mixing appropriate volumes of tritolyl phosphate (n_{20}^{D} 1·556) with either butyl stearate (n_{20}^{D} 1·445) or liquid paraffin (n_{20}^{D} 1·47), giving refractive index steps of 0·001–0·002 within ranges of particular interest, e.g., n 1·47 and n 1·56. For mixtures of higher refractive indices 1-bromonapthalene (n_{20}^{D} 1·658) can be used. The mixed liquids must be calibrated with a refractometer and kept in dark, well-stoppered bottles. The procedure for establishing the fibre refractive index is to choose the liquid of nearest match in refractive index from the series using the Becke-line effect and with the fibre oriented either parallel ($n_{\parallel}$) or perpendicular ($n_{\perp}$) to the vibration direction of plane-polarized light. The Becke-line effect[7] denotes whether a particular liquid in use is higher or lower in refractive index than the fibre. Monochromatic light should be used which can be from either a sodium lamp or a sodium-line interference filter. Generally, mercury green (546 nm) or mercury yellow (578 nm) are used because an intense source can be isolated by interference filters from a high pressure mercury-vapour lamp. If monochromatic light is used other than sodium and precise measurement of refractive index is required, then the refractive index of the liquid will have to be determined for that wavelength of light using a knowledge of the dispersion of refractive index with wavelength.

The measurement of birefringence ($n_{\parallel} - n_{\perp}$) may be made independently of refractive index determination and more simply by the use of a polarizing microscope fitted with either a quartz wedge mounted in a Dick-Wright ocular[8] or a compensator,[9] e.g., a Berek. The fibre is arranged on the microscope stage at 45° (diagonal position) to the planes of polariza-

---

[7]When using the Becke-line method to measure a fibre refractive index, care should be taken if the specimen is a bicomponent fibre, particularly a sheath–core fibre, when misleading values can be obtained.

[8]Obtainable from Vickers Instruments Ltd.

[9]N. H. Hartshorne and A. Stuart. 'Crystals and the Polarizing Microscope', E. Arnold, London, 1966.

tion of the polarizer and analyser which are crossed. It is necessary to determine the thickness of the fibre at the point of measurement, however, and this can present difficulty with fibres of non-circular cross-section. It must be stressed that when measuring fibre diameter the condenser diaphragm must be only slightly stopped down, otherwise the edges of the fibre will appear to have a thick black outline. Unless the fibre outline is seen as a fine dark line, errors in diameter measurement may cause relatively large errors in birefringence determinations, particularly with fibres of low linear density. If the image is too bright, the intensity of illumination should be reduced at the lamp rheostat or by the introduction of neutral density filters in the light path, but not by stopping down the condenser diaphragm.

A rapid method of birefringence measurement, suitable for routine work, uses a quartz wedge compensator with a scale marked directly on the wedge. By arranging that the image of the fibre formed by the objective lens is in the same plane as the wedge, both fibre and wedge may be viewed in focus together and the position of the zero-order fringe read directly off the scale. The fitting of a Dick-Wright ocular to a polarizing microscope allows this to be done. For most fibres, $n_\parallel$ is greater than $n_\perp$ and the wedge compensator should be cut with the fast direction along the length of the wedge so that curved fringes are seen across the fibre as with the Berek Compensator. This is contrary to that normally supplied and must be specified. However, a wedge can be used with the fast direction across the wedge, when the measurement of path difference (retardation) is made by bringing the zero-order fringes (aligned along the fibre length) to meet at the centre of the fibre as with a Senarmont compensator. A Berek compensator is fitted so that compensation of the path difference (retardation) introduced by the fibre can be obtained with white light to permit identification of the zero-order fringe. The path difference can be obtained from the compensator calibration and converted into the birefringence knowing the thickness of the fibre.

Table B3.2 shows refractive indices determined by interference microscopy which indicates the average optical properties through the fibres. The values given may thus differ slightly from the superficial optical properties determined by the Becke-line technique; experience suggests that the difference will not exceed 0·005. Mercury yellow light (578 nm) was used in preference to sodium light (589 nm) and the difference in index expected between the two wavelengths will be less than 0·001.

Results were obtained on fibres that had been previously relaxed in water at 95 °C for 1 min and then air dried at 65% relative humidity for 24 hours. The relaxation treatment reduced the variability of the values from different fibres in the same sample, presumably by removing local strains and provided values more in keeping with those obtained on fibres extracted for test purposes from finished fabrics.

5.5 Optical Test for Poly(ethylene terephthalate)

The test depends on the fact that the refractive index of poly(ethylene terephthalate) for light polarized in a direction parallel to that of the fibre axis is 1·71–1·74, considerably higher than that for light polarized at right angles to the fibre axis, and also higher than the refractive index of most other fibres[10] for light polarized in either direction. When poly(ethylene terephthalate) is immersed in a liquid of refractive index 1·72 it is, therefore, almost invisible in light polarized in a direction parallel to the fibre axis, and visible in light polarized in any other direction, or in normal light. No other fibres become almost invisible in a medium of this refractive index but some confusion could occur with aramid fibres[10].

To prepare a suitable medium mix 1 part of monobromonaphthalene and 3 parts of di-iodo methane. It is advisable to test the mixture against a sample known to be poly(ethylene terephthalate).

5.6 Method for Obtaining Skin-stained Cross-sections of Regenerated Viscose Fibres

Tie a parallel bundle of filaments across a wire frame and dye for 1 hour at the boil in a 4% solution of Chlorazol Blue FF (Colour Index Direct Blue 1); rinse thoroughly and allow to dry in air. Lightly stretch the filaments across an embedding cell; embed in a methacrylate cement using the method described for thin cross-sections (Section 4.1.5.3).

Cut sections of 10 μm thickness and transfer to a dish containing 20% aqueous pyridine; after 3 to 5 min transfer to filter paper and allow to dry. Put the dried sections on a microscope slide in liquid paraffin (n 1·47) and examine. The results are given in Table VIII (g).

5.7 Measurement of Fibre Densities

The density of a fibre has proved to be of value in aiding identification and also, in the case of man-made fibres, in providing an indication of crystallinity caused by heat treatments during processing.

One of the simpler methods of density determination is by the use of a density gradient column containing calibrated glass density floats. The positions or levels of the floats in the column are measured and a calibration chart plotted of float position (cm) against float density (g/cm³), which should give a linear curve. From this calibration chart it is a simple procedure to measure the position of a specimen in the column and determine the corresponding density.

The preparation of a gradient column is very simple. (See Fig. 5.1, which is based on the Tecam unit.)

Connect two Erlenmyer flasks (1 l) A and B by the tap unit T_1 and place the stirrer in flask B. Connect tap T_2 to the capillary tube which is passed through the column cover cap almost to the bottom of the column tube. Make sure taps T_1 and T_2 are closed. Into flask A pour 500 ml of the denser liquid and into flask B pour 500 ml of the less dense liquid. Start the magnetic stirrer and run it at a suitable speed without causing too much

---

[10]Aramid fibres will give a positive result with this test because of their high refractive index for light polarized in a direction parallel to that of the fibre axis (see Table B3.2). They may be readily distinguished from polyester fibres by their different thermal properties (see Table B4. (Appendix B)).

surface fluctuation. Open tap T_1 and allow hydrostatic equilibrium to be established. Open tap T_2 and regulate the flow so that the column will take at least 1½ hours to fill. Do not alter the flow rate again, or the column will be non-linear. When the column has filled to 5 cm from the top close T_2 and T_1, disconnect the capillary tube from T_2 and very carefully and slowly withdraw the tube from the column.

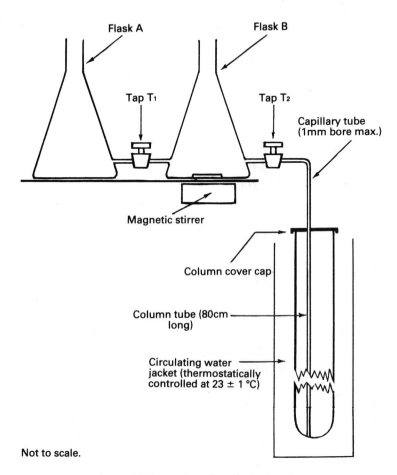

Flask A

Flask B

Tap T₁

Tap T₂

Capillary tube
(1mm bore max.)

Magnetic stirrer

Column cover cap

Column tube (80cm long)

Circulating water jacket (thermostatically controlled at 23 ± 1 °C)

Not to scale.

Diag. 5.1 Preparation of gradient column

Remove the column cover cap and position the wire mesh basket containing the required range of density floats over the column. Switch on the motor so that the basket is lowered into the column. It will take about 30 min for the basket to reach the bottom having left the density floats at the appropriate levels in the column. Cover the column and allow it to stabilize overnight.

The level of the floats is measured using a cathetometer or similar measuring system and the calibration curve is plotted. Provided the curve is linear the column is now ready for use. Although this procedure is based on the Tecam Density Gradient Column apparatus it can be applied to any other similar system.

For fibres known to be homogeneous, a sample is prepared for measurement by taking some fibres and tying them in a small knot of which the loose ends are snipped off close to the knot. The knot is then placed in a test-tube with a few ml of xylene[11] and boiled for about 2 min (*note—flammable*) to dry the fibres and to remove air from the mass. The knot is then picked up by a pair of forceps and immediately dropped into the density gradient tube. After about 30 min the fibres will generally have sunk very near to their final equilibrium positions. The density corresponding to this position can be found quickly by reference to the positions of the calibrated floats.

Mixtures of fibres can be examined by cutting them into short lengths so that the individual fibres separate. These separate fibres are hard to see by ordinary lighting, but can be made easily visible by using crossed polars, respectively, before and behind the tube. When a light is shone through, the double refractions of the fibres will make them show up bright against a dark ground.

Notes
  (*a*) The choice of solutions will depend on the density range required in the column, the accuracy required, and the type of specimens to be tested. Some solutions could dissolve the specimen. The solutions used must be miscible.
  (*b*) Finishes, such as crease-resistant and waterproof, will affect the true density and, if possible, should be removed before carrying out the test.
  (*c*) Specimens can be dredged out by reversing the motor and pulling the basket, floats, and specimens out. The specimens can then be picked out of the basket using tweezers, after which the motor is reversed lowering the basket and floats into the column again.
  (*d*) Depending on the usage a well maintained density column will last for months, but it should be checked and re-calibrated at intervals.

5.8   Melting Points of Fibres

The melting point can be measured conveniently by means of an L-shaped copper block[12]. In use the block is inverted (⌐), the upright part being slotted to hold the fibres and drilled for a thermometer pocket. The shape of the copper block enables the over-hanging part of the inverted L to be heated in a Bunsen flame without the hot gases reaching the fibres.

Fibres are placed in the slot, the block is heated slowly, and the temperature at which the fibres soften can be observed to within quite narrow limits.

In certain cases it is easier to determine the melting point of a fibre immersed in silicone fluid, and the following method is suitable. The fibre is placed at the bottom of a glass tube approximately 10 cm long and 2 mm internal diameter. Silicone Fluid MS 550[13] is added to the tube by means of a dropper until the fibre is totally immersed. The tube is attached to the bulb of a thermometer and the melting point determined by the classical

---

[11]Note that polyolefin and PVC-type fibres dissolve in xylene under these conditions.
[12]J. M Preston. *J. Text. Inst.*, 1949, **40**, T767.
[13]Available from Midland Silicones Ltd.

method using a stirred melting-point bath. Silicone fluid is preferred as the heat transfer medium since liquid paraffin fumes badly above 200 °C.

The softening ranges and melting points set out in Table B4 (Appendix B) have been obtained using these two methods. Where a reasonably sharp melting point exists no figures for softening range are given.

The melting points given in various references or publications for similar fibres may show considerable variation. The factors that affect the melting point are the moisture content of the fibre, whether the fibre is in air, silicone immersed or in a sealed tube, the rate of heating, and the presence of any additives incorporated during manufacture.

It is, therefore, essential that workers should standardize their own method and determine melting points on authentic samples for future reference.

5.9   Preparation and Measurement of Ultimate Fibres

The material under examination is sampled by drawing at random several tufts of fibre. These are bulked and, if necessary, pulled apart and thoroughly mixed to obtain a homogeneous sample; from this a small representative piece weighing approximately 2 g is drawn. If there is any possibility that the sample under examination is ramie, the fibre length of the sample taken should be at least 50 cm.

Treat the sample as follows:

1.   Boil it gently in 1% sodium hydroxide solution for 30 min, filter on a Buchner funnel, and wash it first with hot water and then with a 5% solution of acetic acid.

2.   Steep the sample for 30 min in a solution containing 5 g sodium chlorite and 5 g acetic acid per 100 ml at 95–100 °C, preferably in a fume cupboard since chlorine dioxide is evolved.

3.   After filtering on a Buchner funnel, wash the fibre mass well with hot water and transfer it to a wide-mouthed polythene bottle containing about ten rubber-covered glass rods, each about 2 cm in length and 6 mm in diameter, the dimensions being such that each rod forms a tight fit in a rubber pipette filler. Partly fill the bottle with water at 70–80 °C, stopper it, and shake it vigorously, preferably in a shaking machine.

4.   Filter the ultimate fibres produced on a Buchner funnel, and if the ultimates are well separated wash them with water, and, if not required for immediate staining, etc., store in glycerol. If the separation is not complete, repeat stages 2 and 3.

The ultimates are stained with 1% aqueous Methylene Blue (see Section 5.17.9) and are teased out on a microscope slide so that they are well distributed.

The dimensions of the fibres are most conveniently measured by means of a projection microscope; a millimetre rule may be used for short straight fibres and for measuring diameters, but for long fibres[14] a map measurer is

---

[14]The long ultimates from fibres such as flax, hemp, and ramie, are more conveniently measured by laying them on a glass slide smeared with glycerine and using a millimetre rule (E. Slattery. *J. Text. Inst.*, 1936, **27**, T101).

useful. A mechanical stage is almost indispensable for obtaining controlled movement of the slide. For length measurements a magnification of 50× will be found to be satisfactory and for diameter[15] measurements 300×.

The number of measurements to be made, will, of course, depend on the degree of accuracy required; the figures given in Table B1.2 (Appendix B) are based on 100 measurements of length and 50 measurements of diameter for each sample.

5.10   Test for Distinguishing Between Abaca and Sisal (Billinghame's Test)

The test devised by Swett[16] for the differentiation of abaca (manila) from other fibres used in rope manufacture has certain disadvantages. These are the uneven staining effect, the difficulty in distinguishing at a glance between the colours produced, and the tendency of the colours to fade. The modified method of Billinghame[17] produces both greater colour differences and more stable colours. The method of applying Billinghame's test is as follows.

1.   Wash the sample with methylene chloride to remove oil and allow it to dry.
2.   Boil the sample in 5% nitric acid for 5 to 10 min.
3.   Wash out the excess acid with water and immerse the sample in cold 0·25N sodium hypochlorite for 10 min (see Section 5.17.13).
4.   Remove the sample and dry it.

Abaca (manila) assumes an orange colour, whereas sisal and other leaf fibres are stained a pale yellow colour.

Dave and Mehta[18] have described a similar method for making quantitative estimations.

5.11   Phloroglucinol and Hydrochloric Acid Test

Treat the fibres with a 2% alcoholic solution of phloroglucinol together with an equal volume of conc. hydrochloric acid. Unbleached jute gives a deep magenta colour (due to the presence of lignin). Other vegetable fibres, e.g., hemp and kapok, may also be stained, but the colour produced is much less intense.

---

[15]It will be appreciated that the term 'diameter' cannot be strictly applied to the bast fibres, none of which have a definite diameter. The term here refers to the distance between the outer walls of the fibre as viewed from above, and it is this measurement that is almost invariably quoted in the literature.

[16]C. E. Swett. *Industr. Engng Chem.*, 1918, **10**, 227.

[17]A. V. Billinghame. *Text. Mfr*, 1940, **66**, 118.

[18]R. H. Dave and S. B. Mehta. *Text. Res. J.*, 1963, **33**, 320.

5.12   Drying-twist Test[19]

Principle. If a wet flax fibre is held at one end and allowed to dry, the free end, directed towards the observer, will be seen to move in a clockwise direction. Hemp fibre under similar conditions follows an anticlockwise movement. The test may be applied to material at any stage of manufacture and is successful even with fibres that have been mercerized.

Apparatus   (i) A pair of forceps with fine tips meeting accurately.

(ii) A hot-plate or any source of low temperature heat with a dark background. For this a piece of stout sheet iron about 152 mm square is suitable; it may be heated by a small spirit lamp or gas burner. It is more convenient, however, to use an electrically heated plate; a plate with a dull black surface is desirable for convenience in observing the movements of the fibres.

Conditions of test. When examining the yarn or cloth, in the latter case after separating the warp and weft, cuttings about 25 mm long are taken from various places in the material in such a way as to obtain fair sampling. The cuttings are soaked for a few minutes in water and then, with the aid of forceps, fibres are withdrawn and held over the hot-plate in order to observe the direction of the twist on drying.

Precautions. As far as possible single fibres should be withdrawn and examined. Small bundles of fibres, however, can be examined, provided that sufficient time is given over the hot-plate to ensure that the drying twist has set in.

Other fibres. Cotton fibres, under this test, usually twist alternately in opposite directions in a very irregular manner. After some experience the presence or absence of cotton in a material may be determined with certainty by this test. Ramie and nettle-fibre show a drying twist in the same direction as flax, but most other fibres twist in the same direction as hemp.

5.13   Examination of Fibre Ash

The microscopical examination of the ash of a vegetable fibre often reveals the presence of structures of value in its identification (see bast and leaf fibres, Figs. 79–81). It must be emphasized, however, that, since these crystalline structures are contained in the tissues that adhere to the fibres, they will be scarcer in fibres that have received such processing and, in any case, their absence is not conclusive.

Ashing is conveniently carried out on a piece of mica over a small flame, and the ash is carefully transferred to a microscope slide. As a mountant, water may be used or a permanent preparation made simply by adding Canada balsam and a cover-slip.

The cluster crystals (Figs 79 (*b*), 80 (*a*)), solitary crystals (Fig. 79 (*a*)), and rod-like crystals (Fig. 80 (*b*)) exist in the ash in the form of calcium carbonate, and are readily dissolved by hydrochloric acid. The stegmata (Fig. 81 (*a*) and (*b*)) on the other hand, do not dissolve, and can be further characterized by the fact that they appear red when mounted in a mixture of phenol and clove oil.

---

[19]C. R. Nodder. *J. Text. Inst.,* 1922, **13**, T161.

5.14 Staining Test for Normal Dyeing, Deep-dyeing, and Basic-dyeable Types of Nylon

Make up a solution containing:

Lissamine Blue 4GL	1·0 g
Synacril Yellow 8G	0·5 g
Sodium dihydrogen phosphate	10·0 g
Di-sodium hydrogen orthophosphate	6·0 g
Matexil VL	0·5 ml
Water	500·0 ml.

Immerse the sample in boiling solution for 30 seconds and wash off. The colours obtained are as follows:

Standard-dyeing nylon	green
Deep-dyeing nylon	blue
Basic-dyeable nylon	yellow.

It is essential to compare the unknown sample with reference samples.

5.15 Meldrum's Stain[20] and Sevron Orange Stain

5.15.1 Meldrum's Stain

Make up two solutions containing:

1. Lissamine Green SFS (Colour Index Acid Green 5)	0·5 g
Approximately 2N sulphuric acid	5·0 ml
Water to	500·0 ml
2. Sevron Orange L (Colour Index Basic Orange 24)	0·5 g
Approximately 2N sulphuric acid	5·0 ml
Dispersol VL or equivalent	1·0 ml
Water to	500·0 ml

Mix equal parts of these two solutions as required.

5.15.2 Sevron Orange Stain

Use solution 2 as prepared for Meldrum's stain.

5.16 Solubility and Staining Tests for Acrylics

It is convenient to conduct the first two tests in Table V (*c*) simultaneously.

Place about 2 ml of Meldrum's stain and nitromethane in separate test-tubes, and immerse them in a water bath maintained at 90 °C. When the tubes and contents have reached this temperature, add the undyed acrylic fibre that has been lightly rolled into a fluffy mass of diameter about 3 mm. Shake the test-tubes gently at intervals throughout the test.

---

[20]K. Meldrum. *J. Soc. Dyers Col.*, 1961, **77**, 22.

After 5 min note whether the fibre has dissolved in the nitromethane. Remove the sample from the Meldrum's stain, wash it first in warm, then in cold water, and dry it. Compare the colour of the fibre with the descriptions given in Table V (*c*).

The results obtained from these initial tests should place the specimen in one of six groups comprising the preliminary classification. For final identification apply the additional tests as set out in Table V (*c*) working from left to right.

Apply the Sevron Orange test in a similar manner to that described for Meldrum's stain using a boiling water-bath.

The temperature of dissolution in DMF is the lowest temperature at which the fibre dissolves within 2–3 min and is determined as follows. About 2 ml of DMF in a ½-in test-tube are warmed in a water-bath to a selected temperature, and the specimen is added. If a specimen dissolves within 2–3 min the test is repeated with fresh solvent and specimen at a temperature 5 °C below the first one, and the procedure is repeated at temperatures diminishing by 5 °C steps until the specimen fails to dissolve in the specified time. If the specimen fails to dissolve at the temperature first selected the procedure is repeated (always with fresh reagent and specimen) at temperatures increasing by 5 °C steps, until the temperature of dissolution is reached.

Form–dimethylamide is decomposed by exposure to light even when stored in a brown bottle. Before use ensure either that the reagent is fresh or has been stored away from the light, preferably in a cool place.

5.17 Preparation and Use of Reagents

5.17.1 Alkaline Lead Acetate

Add strong caustic soda solution to a saturated solution of lead acetate until the precipitate at first formed begins to dissolve. Allow the precipitate to settle and decant the clear liquid. Immerse the sample in this solution at room temperature. Wool is the only fibre that stains dark brown. Wash the sample thoroughly in cold water and neutralize it in dilute hydrochloric acid. Wash and dry the sample.

5.17.2 Beilstein Test

Heat a copper wire in a Bunsen flame until any green coloration disappears. Remove the wire from the flame and touch the fibres with the hot end so that some adhere to it. Again introduce the wire into the flame. The presence of chlorine in the fibre is indicated by a green colour in the flame.

5.17.3 Calcium Chloride in Formic Acid

Dissolve 10 g anhydrous calcium chloride in 100 ml of 90% aqueous formic acid. Add the fibre to the solvent at room temperature using a fibre to liquor ratio of approximately 1 : 500, with occasional gentle swirling. The fibre is considered to be insoluble if it remains undissolved after 15 min.

5.17.4   Celliton Blue FFR

Treat sample for 5 min at 60 °C in distilled water containing 2 g/l Celliton Blue FFR. Liquor ratio, 20 ml solution to 1 g fabric. Rinse very thoroughly before examining.

5.17.5   Congo Red[21]

Swell the specimen in 11% caustic soda. Wash it and stain it in saturated Congo Red (Colour Index Direct Red 28) solution for 10 min. Wash the specimen and mount it in 18% caustic soda solution for microscopical examination.

5.17.6   Hydrochloric Acid Solutions

4·4N

To 55 ml of water add 45 ml of 32% hydrochloric acid. Allow the solution to cool and adjust to sp.gr. 1·075 at 20 °C.

5N

To 49 ml of water add 51 ml of 32% hydrochloric acid. Allow the solution to cool and adjust to sp.gr. 1·085 at 20 °C.

5.17.7   Identification Stains

These stains provide a rapid means of distinguishing between undyed fibres. It should be noted that it is not advisable to use these stains with dyed fibres, since the fibres must first be stripped of their original dyes and the chemicals used for this may well affect the colour produced by the stain, thus giving erroneous results.

There are three principal makes of stain on the market. They are the Shirlastains[22], Du Pont Fibre Identification Stain No.4[23], and the Neocarmin Stains[24]. They may all be used on a variety of fibres, but for full details of their correct use the reader is referred to the manufacturer's literature.

5.17.8   Lissamine Red 2G-Solophenyl Blue Green BL

Treat sample for 15 min at 60 °C in the following solution:

Lissamine Red 2G (Colour Index Acid Red1)	0·5 g/l
Solophenyl Blue Green BL	0·5 g/l
Sodium Chloride	2·5 g/l
Glacial Acetic Acid	0·5 ml/l

Liquor ratio, 20 ml solution to 1 g fabric. Rinse the sample very thoroughly before examining.

[21]G. G. Clegg. *J. Text. Inst.,* 1940, **31,** T49.

[22]Shirlastains are supplied by Shirley Developments Ltd, P.O. Box 6, 856 Wilmslow Road, Didsbury, Manchester, M20 8SA, England.

[23]Du Pont Stain No. 4. is supplied by E. I. Du Pont de Nemours & Co. (Inc.), Organic Chemicals Dept, Dyes & Chemicals Division, Wilmington, Delaware 19898, USA.

[24]Neocarmin stains are supplied by Fesago, Chem. Fabrik Dr. Gossler G.m.b.H., Carl-Benz Strasse 18, D-6902 Sandhausen b. Heidelberg 1, W. Germany.

5.17.9 Methylene Blue

Prepare a dilute aqueous suspension of ultimate fibres by taking some of the fibres direct from the Buchner funnel or from glycerol, and add to about 10 ml of water. Add one or two drops of 1% Methylene Blue (Colour Index Basic Blue 9) to the suspension and shake it. The fibres are fully stained after 1–2 min. Transfer some of the suspension to a 76 × 25 mm glass slide using either a wide bore pipette or mounted needles, as appropriate. Distribute the fibres on the slide using mounted needles if necessary. Place a cover-slip gently over the fibres and remove any excess liquid with a filter paper.

5.17.10 Millon's Reagent

Dissolve 2 g of mercury in 2 ml of concentrated nitric acid at room temperature (the operation should be carried out in a fume cupboard). Then add 2 ml of cold water. If a yellow turbidity develops, stir in a few drops of nitric acid until the solution clears. The solution will keep for several months if stored in an air-tight glass bottle. Immerse the fibres in the hot reagent for 2 min. A pink or red colour develops with natural protein fibres; regenerated protein fibres stain brown.

5.17.11 Soda-lime Test

Place a few fibres in a small ignition-tube, and cover them with soda-lime. Place a loose plug of cotton or glass wool in the mouth of the tube to prevent spitting. Heat the tube strongly and test the vapour with moist red litmus paper. If the vapour is strongly alkaline it contains ammonia, and nitrogen is present in the fibre.

5.17.12 Sodium Fusion Test

Put a piece of clean dry sodium, about the size of a small pea, in an ignition-tube and cover with 25–50 mg of finely divided sample. Heat the tube gently so as to melt the sodium and then more vigorously until the bottom of the tube softens. Shatter the tube by plunging it into 10 ml of distilled water, transfer the mixture to a test-tube, boil, and filter it. To a portion of the filtrate add a drop of dilute alkaline lead acetate solution. A black precipitate (or dark coloration) indicates the presence of sulphur in the sample. (N.B., protective goggles or a screen should be used during the sodium fusion test.)

5.17.13 Sodium Hypochlorite Solutions

1N

Commercial sodium hypochlorite is generally received at a strength of 4N or stronger. The exact strength of the solution should be determined by titration of the available chlorine, after which the strength is adjusted to 1N by dilution with the correct volume of distilled water. A normal solution of sodium hypochlorite contains 35·5 g available chlorine per litre. The solution is unstable, but storing in a cool dark cupboard prolongs the effective life of the reagent to some extent.

1N plus 0·5% sodium hydroxide

Add 5 g of sodium hydroxide pellets to 1 l of the 1N solution previously described.

0·25N

Make up 2 ml of normal sodium hypochlorite to 8 ml with distilled water.

5.17.14   Sulphuric Acid Solutions

75% w/w

To make 2 l of this solution add 1360 ml of conc. sulphuric acid slowly to 700 ml of water contained in a round-bottomed flask (*not* in a bottle), cooled under a tap. After the solution has cooled to room temperature, adjust the sp.gr. to 1·67 at 20 °C.

60% w/w

To make 500 ml of this solution slowly add 250 ml of conc. sulphuric acid to 320 ml of water contained in a round-bottomed flask. After cooling the solution to 20 °C, adjust the sp.gr. to 1·4875 (limits 1·485–1·490) by accurate temperature-controlled hydrometry or by specific gravity bottle.

35% w/w

To 200 ml of water slowly add 55 ml of 98% sulphuric acid. Allow it to cool and adjust if necessary to sp.gr. 1·260 at 20 °C.

2N (approximately)

To 94·5 ml of water add slowly 5·5 ml of 98% sulphuric acid and allow the solution to cool.

5.17.15   Trypsin Test

Prepare a solution of 1 g commercial crude trypsin in 100 ml distilled water, and add 0·3 g sodium bicarbonate to bring to pH 8·2. Boil the fibre for 15 min with 0·2% sulphuric acid, and wash it thoroughly. Heat the solution of trypsin to 40 °C and add the fibre to the solution in the ratio of 1 : 500 approximately. Maintain the temperature at 40 °C for 30 min. Merinova dissolves within 30 min; wool, hair, silk, and all non-protein fibres do not.

5.17.16   Zinc Chlor-iodide (Herzberg Stain)

Dissolve 20 g zinc chloride in 10 ml water. Add 2·1 g potassium iodide and 0·1 g iodine dissolved in 5 ml of water. After the precipitate has settled decant the clear liquid; add a leaf of iodine and store the solution in a dark bottle.

The sample of fibre is wetted out well, squeezed and immersed for 3 min in the test solution, and then rinsed in water. Lignified cellulose is deeply stained.

5.18   Examination of Epidermal Tissue

The distinction between hemp and sunn fibre is difficult (see p. 18) and depends on the appearance of the epidermal tissue. The quantity of epidermal tissue present is small in fibre that has received much processing. Epidermal tissue is present as fragments adhering to the fibre and may be removed for examination by shaking and rubbing the fibre or by untwisting yarn. The particles produced are collected and mounted on a slide, a drop of water or liquid paraffin is added, and a cover-slip placed on top.

5.19   Fibre Diameters [25]

A useful conversion formula for determining the linear density of a cylindrical fibre from its diameter is as follows:

$$T = \frac{d^2\rho}{1274}$$

where T = linear density of cylindrical fibre (tex),
d = fibre diameter (μm), and
ρ = fibre density (g/cm^3).

5.20   Staining Method to Show Nylon 6 and Nylon 6.6 Distribution in a Bicomponent Fibre Cross-section

This method is dependent on differential staining between the nylon 6 and nylon 6.6.

Prepare a staining solution of 2·5 g iodine dissolved in a solution of 20 g potassium iodide in 100 ml water.

Cross sections can be prepared by either the plate method or the microtome method (see Sections 4.1.5.1, 4.1.5.2). In the plate method a drop of the iodine stain is placed on the section in the plate and after 15 seconds the excess stain is blotted away using filter paper. Microtome sections are placed in a drop of the stain on a cavity slide and after 15 seconds the sections are removed and the excess stain blotted away. The sections are then mounted on a microscope slide.

Examination of the sections using transmitted light will show the nylon 6 component to be stained a darker colour than the nylon 6.6. The degree of staining will depend on the time the sections are in the stain, but ideally the nylon 6 component should appear dark brown or black and the nylon 6.6 should appear light brown (see Fig. 136).

---

[25]See also 'Finding the Denier of Man-made Fibres from Fibre Diameter Measurements', Wira Report No. 26, 1968.

Q

Appendix A—Quantitative Analysis

It is outside the terms of reference of this book to include comprehensive details of methods for the quantitative analysis of fibre blends. Details of the methods for analysing blends of animal fibres, animal and non-animal fibres, and various man-made fibres are well documented elsewhere, and, therefore, only two microscopical methods are given here.

Blends containing man-made fibres are analysed by a system of sequential dissolution of the different components in the appropriate solvents. Methods for analysing binary and ternary mixtures have been the subject for discussion on the establishment of international standards; some methods have already been accepted. Reference should be made to the national and international standard procedures that have been published[1].

Generally in these methods the major component is dissolved out first, then one of the other components. Because the solvent for one fibre may cause a loss in weight of another component it may be necessary to dissolve the components in a special order and/or to apply a correction factor; all these details can be found in the test method specifications.

A1 Removal of Non-fibrous Matter[2]

It is common practice for various additions to be made to fibres, yarns, and fabrics for the purposes of assisting processing and manufacture or modifying the properties of the finished material. These usually result in an appreciable increase in mass and often affect the solubility of the individual fibres. It has also to be borne in mind that fibres generally contain a small proportion of naturally occurring non-fibrous substances. The removal of all these non-fibrous substances is therefore necessary before conducting the procedures for quantitative chemical analysis. The removal of some non-fibrous matter, particularly when more than one substance is present, may demand the exercise of considerable chemical resource, and each material to be treated for removal of its non-fibrous matter should be regarded as an individual problem. It is not pretended this procedure is complete, and it should not be assumed that the procedures described in Section A1.6 will have no effect on the physical and chemical properties of the textile materials concerned.

It may be assumed that Soxhlet extraction under the conditions described in Section A1.6 will ensure adequate removal of oils, fats, and waxes. With other non-fibrous substances it is necessary, wherever possible, to check that removal is complete. If not, the recommended procedure shall be repeated. If the extraction in light petroleum as described in A1.6.1 is conducted, it is not necessary to repeat this procedure.

---

[1]British Standards Institution, BS Handbook 11; International Standardization Organization, ISO/R 1833; American Association of Textile Chemists and Colourists, AATCC 20A–1971.

[2]Extracts from BS Handbook 11 are reproduced by permission of British Standards Institution, 2 Park Street, London, W1A 2BS.

Precautionary note Since certain hazards are associated with reagents and solvents employed in the methods given below, these methods should be used only by persons acquainted with the hazards and the necessary precautions.

A1.1   Scope

Procedures are described for the removal of certain commonly found types of non-fibrous substances from the fibres. Fibres to which the procedures are applicable and those to which the procedures are not applicable are listed in Section A1.5. Identification of the non-fibrous matter[3] is not covered by this Appendix.

A1.2   Principle

Where possible, non-fibrous matter is removed by a suitable solvent, but in many cases the removal of certain finishes involves some chemical modification of the finish. In addition, chemical degradation of the fibre substance cannot always be avoided.

A1.3   Definition

For the purposes of this method the following definition applies:

Non-fibrous matter This comprises (i) processing aids such as lubricants and sizes (but excludes jute-batching oils) and (ii) naturally occurring non-fibrous substances.

A1.4   Apparatus

The apparatus required is part of the normal equipment of a chemical laboratory.

A1.5   Use of Procedures for Removal of Non-fibrous Matter

See Table A1.

A1.6   Methods for Removal of Non-fibrous Matter

A1.6.1 Oils, Fats, and Waxes (Except Soaking Oils)

Extract the specimen in a Soxhlet apparatus with light petroleum (distilling between 40 °C and 60 °C) for at least 1 h at a minimum rate of six cycles per hour.

A1.6.2 Soaking Oils

Extract the specimen in a Soxhlet apparatus using a mixture of one volume of toluene with three volumes of methanol as the solvent, for at least 2 h at a minimum rate of six cycles per hour.

---

[3]C. H. Giles, and E. Waters, *J. Text. Inst.*, 1951, **42**, p. 909; S. D. Dandekar, and C. H. Giles, *J. Text. Inst.*, 1962, **53**, p. 430.

TABLE A1

Non-fibrous Matter	Procedure	Fibre Mixtures Containing the Following Fibres	
		Applicable	Not Applicable
Oils, fats, and waxes	A1.6.1	Most fibres	Elastanes
Soaking oils	A1.6.2	Nett silk	—
Starch	A1.6.3	Cotton[1], linen[2], viscose, spun silk, jute[3]	—
Locust-bean gum and starch	A1.6.4	Cotton[1], viscose, spun silk	—
Tamarind-seed size	A1.6.5	Cotton[1], viscose	—
Acrylic	A1.6.6	Most fibres[4]	Protein fibres, cellulose acetates
Gelatin	A1.6.7	Most fibres	Protein fibres
Poly(vinyl alcohol)	A1.6.8 (i)	Cellulose acetate	
	A1.6.8(ii)	Polyester	Protein fibres, cellulose acetates
Starch and poly(vinyl alcohol)	A1.6.9	Cotton, polyester	Protein fibres, cellulose acetates
Poly(vinyl acetate)	A1.6.10	Most fibres	Cellulose acetates
Linseed oil sizes	A1.6.11	Viscose crêpe yarns	Protein fibres, cellulose acetates
Amino-formaldehyde resins	A1.6.12	Cellulose, polyester, polyamide	Asbestos
Bitumen, creosote, and tar	A1.6.13	Most fibres	Cellulose acetates, modacrylics
Cellulose ethers	A1.6.14(i)	Cellulose fibres	—
	A1.6.14(ii)	Cotton	Viscose
Cellulose nitrate	A1.6.15	Most fibres	Cellulose acetates
Poly(vinyl chloride)	A1.6.16	Most fibres	Cellulose acetates
Linseed oil	A1.6.17	Cellulosic fibres, polyamide, silk	Cellulose acetates
Oleates	A1.6.18	Most fibres	Cellulose acetates, modacrylics, asbestos
Oxides of chromium, iron, and copper	A1.6.19	Cellulosic fibres	—
Pentachlorophenyl laurate	A1.6.20	Most fibres	—
Polyethylenes	A1.6.21	Most fibres	—
Polyurethanes	A1.6.22	Polyamide	Polyester, cellulose acetates
Rubbers: natural, styrene-butadiene, neoprene, nitrile	A1.6.23	Most cellulosic fibres	All synthetic fibres
Silicones	A1.6.24	Most fibres	—
Tin weighting	A1.6.25	Silk	—
Wax-based waterproof finishes	A1.6.26	Cotton, protein, polyamide	Cellulose acetates

[1]Grey cotton loses mass when treated by these methods. The loss amounts to approximately 3% of the final oven-dry mass.

[2]Linen loses mass when treated by these methods. The amount depends on the types of yarn from which a fabric is woven. Losses are approximately as follows: bleached yarns 2%, boiled yarns 3%, and grey yarns 4%.

[3]Jute loses mass may by approximately 0.5% when treated by this method.

[4]Nylon 6.6 may undergo a loss in mass of fibre substance of up to 1% when treated by this method. The loss in mass of nylon 6 may vary between 1% and 3%.

A1.6.3 Starch

Immerse the specimen in a freshly prepared solution containing 0·1% by mass of a non-ionic wetting agent, together with an appropriate amylase preparation using a liquor:specimen ratio of 100:1. The concentration of the amylase preparation and the pH, temperature, and time of treatment should be those recommended by the manufacturer. Transfer the specimen to boiling water and boil it for 15 min. Test for complete removal of starch using a dilute aqueous solution of iodine in potassium iodide. When all the starch is removed, rinse the specimen thoroughly in water, squeeze or mangle it, and dry it.

A1.6.4 Locust-bean Gum and Starch

Boil the specimen in water for 5 min, using a liquor:specimen ratio of 100:1. Repeat this procedure with a fresh portion of water. Follow this by the procedure described in A1.6.3.

A1.6.5 Tamarind-seed Size

Boil the specimen in water for 5 min, using a liquor:specimen ratio of 100:1. Repeat this procedure with a fresh portion of water.

Note Size prepared from coarsely ground undecorticated tamarind-seed powder may not be completely removed by this procedure.

A1.6.6 Acrylic Size

Immerse and agitate the specimen for 30 min in at least 100 times its own mass of a solution containing 2 g/l soap and 2 g/l sodium hydroxide at 70 °C to 75 °C. Give three 5 min rinses in distilled water at 85 °C, squeeze, mangle, or centrifuge, and dry the specimen.

A1.6.7 Gelatin

Treat the specimen in a solution (using a minimum liquor:specimen ratio of 100:1) containing:

1 g/l of non-ionic surfactant,
1 g/l of anionic surfactant, and
1 ml/l of ammonia (relative density 0·880)

for 90 min at 50 °C followed by 90 min in the same bath at 70 °C to 75 °C. Wash the specimen and dry it.

A1.6.8 Poly(vinyl alcohol)

(i) Conduct the procedure described in A1.6.7.
(ii) Then treat the specimen in a solution (using a minimum liquor:specimen ratio of 100:1) containing:

1 g/l of non-ionic surfactant,
1 g/l of anionic surfactant, and
1 g/l of anhydrous sodium carbonate

for 90 min at 50 °C followed by 90 min in the same bath at 70 °C to 75 °C. Wash the specimen and dry it.

A1.6.9 Starch and Poly(vinyl alcohol)

Conduct the procedure described in A1.6.3, followed by the procedure described in A1.6.8(ii).

A1.6.10 Poly(vinyl acetate)

Extract the specimen in a Soxhlet apparatus with acetone for at least 3 h at a minimum rate of six cycles per hour.

A1.6.11 Linseed Oil Sizes

Conduct the procedure described in A1.6.1, followed by the procedure described either in A1.6.7 or A1.6.8(ii).

A1.6.12 Amino–formaldehyde Resins

Extract the specimen with boiling 0·02N hydrochloric acid for 10 min using a liquor:specimen ratio of 100:1. Wash the specimen in water, drain, wash it in a 0·1% sodium bicarbonate solution, and finally wash it thoroughly in water.

Note This method causes some damage to cellulosic fibres.

A1.6.13 Bitumen, Creosote, and Tar

Extract the specimen with dichloromethane (methylene chloride) in a Soxhlet apparatus. The duration of treatment depends on the amount of non-fibrous matter present, and it may be necessary to renew the solvent.

Note Extraction of jute with dichloromethane will remove also the batching oil, which may be present to the extent of 5% or more.

A1.6.14 Cellulose Ethers

Methyl Celluloses Soluble in Cold (but not Hot) Water. Soak the specimen in cold water for 2 h. Rinse the specimen repeatedly in cold water with vigorous squeezing.

Cellulose Ethers Insoluble in Water but Soluble in Alkali. Immerse the specimen in a solution containing approximately 175 g/l sodium hydroxide at room temperature, or cooled to a temperature of approximately 5 °C to 10 °C for 30 min. Then work the specimen thoroughly in a fresh portion of reagent, rinse it well in water, neutralize it with approximately 0·1N acetic acid, rinse it again in water, and dry it.

A1.6.15 Cellulose Nitrate

Immerse the specimen in acetone at room temperature for 1 h using a liquor:specimen ratio of 100:1. Drain, wash the specimen in three portions of fresh acetone, and allow the entrained solvent to evaporate.

A1.6.16 Poly(vinyl chloride)

Immerse the specimen in tetrahydrofuran at room temperature for 1 h using a liquor:specimen ratio of 100:1. If necessary, scrape off the softened PVC. Drain, wash the specimen in three portions of fresh tetrahydrofuran, drain and allow the entrained solvent to evaporate.

Caution Because of the risk of explosion, tetrahydrofuran should not be recovered by distillation.

A1.6.17 Linseed Oiled Fabrics

Immerse the specimen in a cold solution containing 10 g/l sodium hydroxide until the oil coating has largely dissolved. Boil the specimen in three separate portions of water until the resultant soaps are removed. Squeeze, mangle, or centrifuge, and dry the specimen.

A1.6.18 Oleates

Immerse the specimen in approximately N/5 hydrochloric acid at 30 °C to 35 °C until it is thoroughly wetted. Wash the specimen well and dry it. Extract the specimen in a Soxhlet apparatus with dichloromethane (methylene chloride) for 1 h at a minimum rate of six cycles per hour.

A1.6.19 Oxides of Chromium, Iron, and Copper

Note This method is not applicable if dyes containing chromium have been applied to the material under test.

Immerse the specimen in a solution containing 14 g/l hydrated oxalic acid at 80 °C for 15 min using a liquor:specimen ratio of 100:1. Wash it thoroughly (any copper present will remain as the colourless oxalate; remove this with 1% acetic acid at 40 °C for 15 min and wash the specimen). Neutralize the specimen with ammonia and wash it thoroughly in water. Squeeze, mangle or centrifuge, and dry it.

A1.6.20 Pentachlorophenyl Laurate (PCPL)

Extract the specimen in a Soxhlet apparatus with toluene for 4 h at a minimum rate of six cycles per hour.

A1.6.21 Polyethylenes

Extract the specimen in boiling toluene under reflux.

A1.6.22 Polyurethanes

Some polyurethanes can be removed by solution in dimethyl sulphoxide or dichloromethane (methylene chloride), and subsequent repeated washing of the sample with fresh quantities of solvent. When the fibre composition of the specimen permits, some polyurethanes can be removed by hydrolysis in an aqueous solution containing 50 g/l sodium hydroxide at the boil. Alternatively, an aqueous solution containing 50 g/l sodium hydroxide and 100 g/l ethanol may be used at a temperature of 50 °C upwards.

A1.6.23 Rubbers: Natural, Styrene-butadiene, Neoprene, Nitrile, and Most Other Synthetic Rubbers

Soak the specimen in a hot volatile solvent which swells it considerably (e.g., toluene), and, when it is fully swollen, remove as much of the rubber as possible by scraping. It may be possible in some cases, where the textile fibres are exposed, to wet only the rubber–textile interface and to strip the rubber and textile layers apart almost at once. Continue by boiling the residual specimen with constant stirring in 50 or more times its mass of molten *p*-dichlorobenzene; use a flat-bottomed flask with an attached wide-bore condenser (to allow adequate access of air), and preferably a magnetic stirrer and hotplate.

After 45 min add 1 part 70% tertiary-butyl hydroperoxide per 4 parts *p*-dichlorobenzene present. Boil until decomposition of the rubber is complete (2 h is an average time). Cool the flask to about 60 °C and add an equal volume of toluene. Filter and wash the textile component repeatedly in warm toluene.

Nitrile rubber (i.e., acrylonitrile–butadiene rubber) may require the addition of the same volume of nitrotoluene as of t-butyl hydroperoxide to speed up the dissolution process.

Note 1 Natural rubber should dissolve after being boiled in toluene alone for several hours in the presence of air. Solution may also be effected by heating in diphenyl ether of 150 °C to 160 °C for 2 h and then washing the specimen in toluene.

Note 2 The above treatments are strongly oxidative in character and the properties of the textile material may be affected appreciably.

A1.6.24 Silicones

Scour the specimen in a solution containing 50 ml to 60 ml per litre 40% hydrofluoric acid in a polyethylene vessel at 65 °C for 45 min. Thoroughly wash the specimen, neutralize it, and scour it in a solution containing 2 g/l soap at 60 °C for 1 h.

A1.6.25 Tin Weighting (Silk)

Immerse the specimen in N/2 hydrofluoric acid in a polyethylene vessel at 55 °C for 20 min and stir occasionally. Rinse in warm water. Immerse the specimen in 2% solution of sodium carbonate at 55 °C for 20 min. Wash the specimen in warm water, squeeze, mangle or centrifuge, and dry it.

A1.6.26 Wax-based Waterproof Finishes

Extract the specimen in a Soxhlet apparatus with dichloromethane (methylene chloride) for at least 3 h. Then, to remove any metallic complexes, scour the specimen in a solution containing 10 g/l formic acid and 5 g/l soap at 80 °C for 15 min. Wash the specimen thoroughly in water until it is free from acid.

A2   Point Counting Technique[4]

A new microscopical method for blend analysis was developed a few years ago by the Shirley Institute. The technique involves the evaluation of a cross-section of a bundle of yarns under a high-power microscope that is fitted with a special eyepiece graticule, such as a lattice-type graticule[5] or a

---

[4]B. Lomas. *J. Text. Inst.*, 1975, **66**, 307.
[5]Obtainable from Carl Zeiss (Oberkochen) Ltd, 31 Foley Street, London W1.

modified Chalkley Point Array graticule[6]. The advantage of this particular method is that it gives a direct count of the fibre-blend ratio in terms of volume, which can be easily converted into a percentage weight. It is not necessary to know the dtex of the individual component fibres.

A2.1 Preparation of Yarn Cross-sections

A cross-section of the yarn is prepared using either the plate technique, or by making microtome or ground sections, after first embedding the yarns in a suitable resin. For most purposes the plate technique is adequate and it is certainly the quickest method. It is essential to use good sections as otherwise ambiguity in the final results may occur. It is also necessary to be able visually to identify the various components in the blend. If there is any difficulty differential staining of the components may be of assistance, but, if ambiguity remains even after staining, blend analysis by the point counting method must not be attempted.

A2.2 Counting the Fibre Cross-sections

A point-counting graticule fitted to a microscope capable of magnifying up to ×1000 is used to count the cross-section. The graticule contains twenty-five points that are arranged either as a lattice or in a random pattern as shown in Diag. A1.

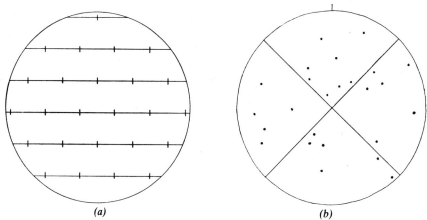

(a)   (b)

Diag. A1. Point-counting graticules.
(a) Lattice type. (b) Random array type

The observer sees a field of view similar to that shown in Diag. A2. The part of the section seen through the graticule is known as the counting field, and the points on the graticule are used as the counting units. Whenever one or more points lies above a fibre cross-section they are counted as one or more hits on that particular fibre component of the blend; if a point lies between the fibres it is not recorded. Several fields are counted until the total number of hits is between 800 and 1000. This method gives good agreement with chemical methods of blend analysis, yet should take an experienced operator only approximately 30 min to carry out.

---

[6]Obtainable from Graticules Ltd, Sovereign Way, Tonbridge, Kent.

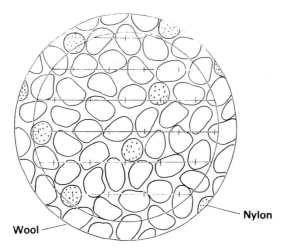

Diag. A2. Counting field seen through graticule.

A2.3 Calculation of Blend Composition

For a two-component blend comprising fibres A and B, with densities ρ_A and ρ_B respectively, where N is the number of hits recorded, conversion of the proportion by volume to the proportion by weight for each fibre component is accomplished as follows:

total number of hits recorded for component $A = N_A$,
total number of hits recorded for component $B = N_B$.

Weight (%) of component A in the blend = W_A

$$\left(\frac{N_A \cdot \rho_A}{N_A \cdot \rho_A + N_B \cdot \rho_B}\right) \times 100,$$

weight (%) of component B in the blend = $100 - W_A$.

Very good correlation between the point counting technique and standard chemical analysis has been obtained. The technique is especially useful if the fabric has a finish that is difficult or impossible to remove, or if time is an important factor.

A3   Quantitative Analysis of Animal-fibre Blends

Wildman[7] has described a method for analysing blends of animal fibres, which is dependent upon the ability of the analyst to recognize microscopically the different types of animal fibres.

Briefly the method involves the determination of the mean fibre diameter of each type of fibre in the blend using a projection microscope as described in IWTO 8–61. The general formula for calculating the weight percentage of the two components A and B in a mixture is:

[7]A. B. Wildman. 'Microscopy of Animal Textile Fibres', Wira, Leeds, 1954.

$$A\% = \frac{n_A(d_A{}^2 + \sigma_A{}^2) \times 100}{n_A(d_A{}^2 + \sigma_A{}^2) + n_B(d_B{}^2 + \sigma_B{}^2)} \, ,$$

where n is the number of fibres measured and counted,
    d is the mean fibre diameter, and
    σ is the standard deviation.

Tables to determine the number of fibres to measure and to count to arrive at a given accuracy are also given by Wildman.

Appendix B—Tables of Data

B1   Fibre Dimensions

B1.1   Dimensions of Some Animal Fibres

TABLE B1.1

Fibre	Approximate Length (mm)	Approximate Fibre Diameter (μm)	
		Range	Mean (or Range of Means)
WOOL			
Merinos	35–90		18–27
Fine Crossbreds	50–100		28–31
Medium Crossbreds	100–200		32–39
Coarse Crossbreds	up to 300		40–44
MOHAIR			
Kid	100–150		25–26
Adult	up to 250		30–55
CASHMERE			
Undercoat (down)	25–75		13–16
Outercoat (Chinese)	75–100	38–142	79
Outercoat (Persian)	75–100	42–160	86
CAMEL HAIR			
Undercoat	25–125	5–40	20
Outercoat	up to 375	up to 120	—
LLAMA			
Alpaca	100–300	10–75	26–27
Llama	250–300	10–150	28–30
Vicuña	35–75	6–35	13–14
HORSE HAIR			
Tail	—	113–233	149–183
Mane	—	73–173	121–129
COW HAIR			
Tail	—	up to 200	—
Body	12–50	12–180	36

The data in Tables B1.1 and B1.2 have been derived from different sources and by different methods, and are, therefore, not all expressed in the same terms. The figures described as 'range of individual fibres' refer to the dimensions observed for such individuals. 'Range of means' refers to data which are individually the means of samples. The latter range is, therefore, much smaller than the former.

B1.2   *Dimensions of Some Vegetable Fibres*

TABLE B1.2

| Fibre | | Length (mm) | | Diameter (μm) |
Common Name	Botanical Name	Range of Means	Range of Maximum Values	Range of Means
COTTON	*Gossypium*			
Indian		12–20	20–36	14·5–22
American		16–30	24–48	13·5–17
Egyptian		20–32	36–52	12·0–14·5
Sea Island		28–36	50–64	11·5–13

| Fibre | | Length of Ultimates (mm) Range of Means | Diameter of Ultimates (μm) Range of Means |
Common Name	Botanical Name		
ABACA, MANILA	*Musa textilis*	4·6–5·2	17·0–21·4
COIR	*Cocos nucifera*	0·9–1·2	16·2–19·5
FLAX	*Linum usitatissimum*	27·4–36·1	17·8–21·6
HEMP, TRUE	*Cannabis sativa*	8·3–14·1	17·0–22·8
JUTE	*Corchorus capsularis*	1·9–2·4	16·6–20·7
JUTE	*Corchorus olitorius*	2·3–3·2	15·9–18·8
KENAF	*Hibiscus cannabinus*	2·0–2·7	17·7–21·9
PHORMIUM	*Phormium tenax*	5·0–5·7	15·4–16·4
RAMIE	*Boehmeria nivea*	125–126	28·1–35·0
ROSELLE	*Hibiscus sabdariffa*	2·6–3·3	18·5–20·0
SISAL	*Agave sisalana*	1·8–3·1	18·3–23·7
SUNN FIBRE	*Crotalaria juncea*	3·7–6·0	23·0–35·0
URENA	*Urena lobata*	2·1–3·6	15·6–16·0

B2 Fibre Densities

TABLE B2

Fibre	Density (g/cm³)	Fibre	Density (g/cm³)
ACETATE		**POLYOLEFIN**	
Diacetate (Dicel)	1·33	Polypropylene (Ulstron)	0·90
Triacetate (Tricel)	1·30	Polyethlene, low-density	
		(Courlene)	0·92
ACRYLIC		Polyethylene, high-density	
Orlon	1·14–1·17	(Courlene X3)	0·95
Courtelle	1·14–1·18		
Zefran	1·15	**POLYURETHANE ELASTOMERS**	
Acrilan, Creslan	1·17	(approximate values)	
		Enkaswing, Lycra, Sarlane,	
ALGINATE		Spanzelle	1·1
Calcium alginate	1·72		
		REGENERATED PROTEIN FIBRE	
CHLOROFIBRE		Merinova	1·29
Heat-treated PVC (Isovyl)	1·38		
Unmodified PVC (Rhovyl,		**VINYLAL**	
Fibravyl)	1·40	Kuralon	1·30
Chlorinated PVC (Piviacid)	1·54		
Polyvinylidene chloride)		**VISCOSE**	1·51–1·52
(Velan, Saran)	1·70		
		RABBIT	
CUPRO		Common	0·92
Cuprammonium rayon	1·52	Angora	1·10
FLUOROFIBRE		**WOOL**	
Teflon	2·30	Non-medullated	1·31
GLASS		**SILK**	
E-glass	2·53	*B. mori,* raw	1·33
C-glass	2·46	Weighted	>1·60
A-glass	2·46	Tussah	1·32
S-glass	2·45		
		COTTON	
MODACRYLIC		Scoured	1·55
Dynel	1·31	Mercerized	1·54
Teklan	1·34	Acetylated	1·40–1·50[1]
Verel	1·37		
		FLAX	1·50
NYLON			
Nylon 11 (Rilsan)	1·10	**HEMP**	1·50
Nylon 6 (Perlon)	1·13		
Nylon 6.6 (ICI nylon)	1·14	**JUTE**	1·50
Nomex	1·38		
		RAMIE	1·55
POLYESTER			
Kodel	1·22	**ASBESTOS**	2·10–2·80
Terylene, Vycron	1·38		

[1]According to degree of acetylation

B3   Refractive Indices

B3.1   *Refractive Indices of Mountants for Microscopy*

TABLE B3.1

Mountant	Refractive Index (n_{20}^D)
Water	1·33
n-Heptane	1·39
Silicone fluid 200/100,000 cs	1·406
n-Decane	1·41
Butyl stearate	1·445
Glycerine jelly	1·45
Liquid paraffin	1·47
Olive oil	1·48
Cedarwood oil[1]	1·513–1·519
Anisole	1·515
ALP 1 immersion oil	1·524
Ethyl salicylate	1·525
Canada balsam	1·53
2-Phenylethyl alcohol	1·533
Methyl salicylate	1·537
o-Dichlorobenzene	1·549
Tritolyl phosphate[2] (Tricresyl phosphate)	1·556
Bromobenzene	1·560
1-Bromonaphthalene	1·658
Di-iodo methane (Methylene iodide)	1·74

[1]The refractive index of cedarwood oil changes slightly with time.
[2]Toxic.

B3.2 Refractive Indices of Fibres

TABLE B3.2

Fibre	$n_{\parallel}$	$n_{\perp}$	Birefringence Δn
ACETATE			
Diacetate (Dicel)	1·476	1·473	0·003
Triacetate (Tricel)	1·469	1·469	0
ACRYLIC			
Acrilan 36	1·511	1·514	−0·003
Courtelle	1·511	1·514	−0·003
Orlon 42	1·511	1·515	−0·004
ARAMID			
Kevlar	>2·000		
ASBESTOS			
Chrysotile	1·50–1·56	—	varies
Amosite	1·64–1·69	—	varies
Crocidolite	1·68–1·71	—	varies
CHLOROFIBRE			
Fibravyl	1·541	1·536	0·005
CUPRO			
Cuprammonium rayon (Bemberg)	1·553	1·519	0·034
GLASS			
A-glass	1·542	—	—
E-glass	1·550	—	—
S-glass	1·523	—	—
C-glass	1·541	—	—
MODACRYLIC			
Dynel (unrelaxed)	1·535	1·533	0·002
Teklan	1·520	1·516	0·004
NYLON			
Nylon 11 (Rilsan)	1·553	1·507	0·046
Nylon 6 (Enkalon)	1·575	1·526	0·049
Nylon 6.6 (ICI nylon)	1·578	1·522	0·056
POLYESTER			
Terylene	1·706	1·546	0·160
POLYOLEFIN			
Polypropylene (Ulstron)	1·530	1·496	0·034
Polyethylene (Courlene X3)	1·574	1·522	0·052
VISCOSE			
Normal-tenacity viscose (Courtaulds Triple 'A')	1·542	1·520	0·022
High-tenacity viscose (Tenasco Super 105)	1·544	1·505	0·039
High-wet-modulus viscose (Vincel)	1·551	1·513	0·038
WOOL	1·557	1·547	0·010
COTTON			
Tanguis	1·577	1·529	0·048
SILK			
Degummed	1·591	1·538	0·053
FLAX	1·58–1·60	1·52–1·53	0·06

R

B4 Melting Points of Fibres

TABLE B4

Fibre	Melting Point (°C)
ACETATE	
Diacetate	250–255
Triacetate	290–300
ACRYLIC	Do not melt; decompose with discoloration
ARAMID	Do not melt
CHLOROFIBRE	
Clevyl T	185–190
Isovyl	210–212
Leavil	Does not melt; decomposes with discoloration
FLUOROFIBRES	
Teflon FEP	285
Teflon TFE	Does not melt; decomposes slowly
MODACRYLIC	
Dynel 180	190
Others	Do not melt; decompose with discoloration
NYLON	
Nylon 11	182–186
Nylon 6	210–216
Nylon 6.6	252–260
POLYESTER	
Kodel II	290
Others	250–260
POLYOLEFIN	
Ulstron	165–175
Courlene	108–113
Courlene X3	135
VINYLAL	
Mewlon	232

B5 Conventional Moisture Regain Factors
TABLE B5[1]

Fibre	Regain (%)	Fibre	Regain (%)
WOOL AND ANIMAL HAIR		**VISCOSE**	13·00
Combed fibres	18·25		
Carded fibres	17·00	**ACRYLIC**	2·00
ANIMAL HAIR		**CHLOROFIBRE**	2·00
Combed fibres	18·25		
Carded fibres	17·00	**FLUOROFIBRE**	0·00
HORSEHAIR		**MODACRYLIC**	2·00
Combed fibres	16·00		
Carded fibres	15·00	**NYLON (6.6)**	
		Staple	6·25
SILK	11·00	Filament	5·75
COTTON		**NYLON 6**	
Normal fibres	8·50	Staple	6·25
Mercerized fibres	10·50	Filament	5·75
KAPOK	10·90	**NYLON 11**	
		Staple	3·50
FLAX	12·00	Filament	3·50
HEMP	12·00	**POLYESTER**	
		Staple	1·50
JUTE	17·00	Filament	3·00
ABACA	14·00	**POLYETHYLENE**	1·50
ALFA	14·00	**POLYPROPYLENE**	2·00
COIR	13·00	**POLYCARBAMIDE**	2·00
BROOM	14·00	**POLYURETHANE**	
		Staple	3·50
KENAF	17·00	Filament	3·00
RAMIE (bleached fibre)	8·50	**VINYLAL**	5·00
SISAL	14·00	**TRIVINYL**	3·00
SUNN	12·00	**ELASTODIENE**	1·00
HENEQUEN	14·00	**ELASTANE**	1·50
MAGUEY	14·00	**GLASS FIBRE**	
		Filament with diameter $>5\mu$	2·00
ACETATE	9·00	Filament with diameter $\lessgtr 5\mu$	3·00
ALGINATE	20·00	**METAL FIBRE**	2·00
CUPRO	13·00	**METALLIZED FIBRE**	2·00
MODAL	13·00	**ASBESTOS**	2·00
PROTEIN	17·00	**PAPER YARN**	13·75
TRIACETATE	7·00		

[1]Statutory Instrument 2124: 1973.

Appendix C—SI Units and Conversion Factors[1]

C1   SI Units and Conversion Factors for Mill and Commercial Transactions
TABLE C1

Quantity	SI Units and Their Appropriate Decimal Multiples	Unit Symbol	To Convert to SI Units Multiply Value in Unit Given by Factor Below	
			Unit	Factor
Length	millimetre	mm	inch	25·40
	centimetre	cm	inch	2·540
	metre	m	yard	0·9144
Width	millimetre	mm	inch	25·40
	centimetre	cm	inch	2·540
	metre	m	yard	0·9144
Area	square metre	m²	yd²	0·8361
Volume	litre	l	pint	0·5682
			gallon	4·546
Mass	kilogram	kg	lb	0·4536
	tonne	t	ton	0·9842
Thickness	millimetre	mm	inch	25·40
Linear density	tex*	tex	—	—
	millitex	mtex	—	—
	decitex	dtex	—	—
	kilotex	ktex	—	—
Threads in cloth				
Length	number per centimetre†	picks/cm	picks/inch	0·3937
Width	number per centimetre†	ends/cm	ends/inch	0·3937
Warp threads in loom	number per centimetre	ends/cm	ends/inch	0·3937
Stitch length	millimetre	mm	inch	25·40
Courses per unit length	number per centimetre	courses/cm	courses/inch	0·3937
Wales per unit length	number per centimetre	wales/cm	wales/inch	0·3937
Mass per unit area	gram per square metre	g/m²	oz/yd²	33·91
Twist	turns per metre‡	turns/m	turns/in	39·37

*The Tex System is fully described in BS 947. It is based on the principle that linear density in tex expresses the mass in grams of one kilometre of yarn. Hence, millitex, decitex, and kilotex express the mass in mg, dg, and kg of one km. BS 947 gives conversion factors for all the recognized yarn count systems. The rounding procedure in Appendix A of BS 947 is superseded by BS 4985 which is based on an ISO standard, which gives rounded Tex System equivalents for yarn counts in the six main counting systems.

†For particularly coarse fabrics, the unit 'threads/10 cm' may be used if there is no possibility of confusion.

‡In some sectors, twist is expressed as turns/cm.

[1]Extracts from PD 6469: 1973, 'Recommendations for programming metrication in the textile industry' are reproduced by permission of British Standards Institution, 2 Park Street, London, W1A 2BS.

C2 SI Units and Conversion Factors for Laboratory Use

TABLE C2

Quantity	SI Units and Their Appropriate Decimal Multiples	Unit Symbol	To Convert to SI Units Multiply Value in Unit Given by Factor Below	
			Unit	Factor
Diameter	micrometre	μm	1/1000 in	25·4
	millimetre	mm	inch	25·4
	centimetre	cm	inch	2·54
Cover factor				
Woven fabrics	threads per centimetre × $\sqrt{tex}$ × 10^{-2}	$\dfrac{(threads/cm)\sqrt{tex}}{100}$	$\dfrac{threads/in}{\sqrt{cotton\ count}}$	0·0957
Weft-knitted fabrics	$\sqrt{tex}$ divided by stitch length in mm	$\dfrac{\sqrt{tex}}{stitch\ length\ (mm)}$	$\dfrac{1}{stitch\ length\ (in)\ \times\ \sqrt{worsted\ count}}$	1·172
Twist factor (or multiplier)	turns per metre × $\sqrt{tex}$ × 10^{-2}	$\dfrac{(turns/m)\ \sqrt{tex}}{100}$	$\dfrac{turns/in}{\sqrt{cotton\ count}}$	9·57
Breaking load	millinewton	mN	gf	9·81
	newton*	N	lbf	4·45
	decanewton	daN	kgf	0·98
Tearing strength	newton*	N	lbf	4·45
Tenacity	millinewton per tex	mN/tex	gf/den	88·3
Bursting pressure	kilonewton per square metre†	kN/m²	lbf/in²	6·89

*The newton is the SI unit of force. It is that force which, when applied to a body having a mass of one kilogram, gives it an acceleration of one metre per second squared. For practical purposes during the changeover period, it is sometimes sufficient to use the equation 1 daN ≃ 1 kgf. This is accurate to within 2% but it should be noted that the kgf will be illegal under the EEC Directive on units after 31 December 1977. Full details of the International System of Units are given in BS 3763 and in PD 5686, both of which are concordant with ISO International Standard 1000.

†The N/m² is known as the pascal (Pa). Thus, this unit may be expressed as kPa.

Note. Metrication makes it imperative to avoid the practice of quoting numerical values without reference to units. For example, the expression '20/2' as a yarn designation is either meaningless or grossly misleading in a metric context.

Appendix D—Bibliography

H. M. Appleyard. 'Guide to the Identification of Animal Fibres', W.I.R.A., Leeds, 1960.

H. M. Appleyard. 'Identification of Biological Damage to Fibres', Wira Report 164, 1972.

H. M. Appleyard. 'Identification of Chemical Damage to Fibres', Wira Report 131, 1971.

H. M. Appleyard. 'Identification of Mechanical Damage to Fibres', Wira Report 178, 1972.

H. M. Appleyard. 'Identification of Skin Wool and Shed Fibres', Wira Report 68, 1969.

R. Barer. 'Lecture Notes on the Use of the Microscope', Blackwell Scientific Publications, Oxford.

J. C. Brown. 'The Determination of Damage to Wool Fibres', *J. Soc. Dyers Col.*, 1959, **75**, 11.

H. Brunner and B. Cowan. 'The Identification of Mammalian Hair', Inkarta, Victoria, 1974.

C. H. Carpenter, L. Leney, H. A. Core, W. A. Cote, and A. C. Day. 'Papermaking Fibres', State University College of Forestry at Syracuse University, New York, 1963.

I. E. de Gruy, J. H. Carra, and W. R. Goynes. 'The Fine Structure of Cotton: An Atlas of Cotton Microscopy', Marcel Dekker, New York, 1973.

H. Driesch. 'Welche Chemiefaser ist Das?', Franckh'sche Verlagshandlung, Stuttgart, 1962.

W. Garner. 'Textile Laboratory Manual', Heywood Books, London, 1966.

J. Gordon Cook. 'Handbook of Polyolefin Fibres', Merrow, Watford, 1967.

J. Gordon Cook. 'Handbook of Textile Fibres', Vols. I & II, Merrow, Watford, 4th edition, 1968.

P. Gray. 'Encyclopedia of Microscopy and Microtechnique', Van Nostrand, Princeton, N.J., 1974.

A. F. Halimond. 'The Polarizing Microscope', Vickers Ltd, 1970.

J. W. S. Hearle, J. T. Sparrow, and P. M. Cross. 'Use of the Scanning Electron Microscope', Pergamon Press, London, 1972.

P. Koch. 'Microscopic and Chemical Testing of Textiles', Chapman & Hall, London, 1963.

B. Luniak. 'The Identification of Textile Fibres', Pitman, London, 1953.

W. C. McCrone and J. G. Delly. 'The Particle Atlas', Ann Arbor Science Publishers Ltd, Ann Arbor, Michigan, 1974.

R. T. O'Conner. 'Instrumental Analysis of Cotton Cellulose and Modified Cotton Cellulose', Marcel Dekker, New York, 1972.

J. L. Stoves. 'Fibre Microscopy', National Trade Press, London, 1957.

G. G. Taylor and J. C. Brown. 'The Uses of Microscopy in Textile Dyeing and Finishing', *J. Soc. Dyers Col*, 1953, **69**, 396.

A. B. Wildman. 'The Microscopy of Animal Textile Fibres', W.I.R.A., Leeds, 1954.

T. Zylinski. 'Textile Metrology', Part 1, Wydawnictwo Przemslu Lekkiego i Spozywczego, Warsaw, 1956.

General Index

In the index roman numerals refer to tables in the Scheme of Analysis, numerals prefixed by a letter refer to tables in the Appendices, and italic numerals refer to figure numbers; all other numerals refer to page numbers, e.g.,

> 9—page 9,
> III—Table III, Scheme of Analysis,
> B4—Table B4, Appendix B,
> *17*—Figure 17.

Fibre Index

In the index roman numerals refer to tables in the Scheme of Analysis, numerals prefixed by a letter refer to tables in the Appendices, and italic numerals refer to figure numbers; all other numerals refer to page numbers, e.g.,